RO
3 1

D1243254

CRAFT
BEER
FOR THE
GEEKS

BREWDOG
BREWDOG

BREWDOG

CRAFT BEER FOR THE GEEKS

RICHARD TAYLOR
WITH JAMES WATT & MARTIN DICKIE

First published in the United States of America in 2020 by
Chicago Review Press Incorporated
814 North Franklin Street
Chicago, Illinois 60610

ISBN 978-1-64160-456-7

Printed and bound in China

10 9 8 7 6 5 4 3 2 1

Group publishing director Denise Bates
Senior editor Pauline Bache
Creative director (for Octopus) Jonathan Christie
Creative director (for BrewDog) Simon Shaw
Designer Jeremy Tilston at The Oak Studio
Special photography Paul Winch-Furness and Richard Clatworthy
Food styling Kat Mead
Senior production manager Katherine Hockley

CONTENTS

INTRODUCTION 7

THE POINT OF DIFFERENCE 11

Quality Begins (and Ends) at Home / The Fantastic Four 2.0 / Being Spontaneous / The Science of Flavor / Quality & Freshness

CULTURE CLUB 47

The Culture of Craft / The Charlatans / The Pioneers

THE STYLE COUNCIL 81

Changing Times / Drink Now. Drink Next.

DOG EAT DOG 2.0 117

Beer & Food Pairing

DIY DOG 171

Beer Perfection Begins at Home / Off Flavors / Home-Brew How to...

HOME BREW: FURTHER READING 217

GLOSSARY 219

INDEX 220

BREWDOG

LAGER
BREWDOG
CREW

INTRODUCTION

CRAFT BEER FOR THE GEEKS

Craft beer is a cult of the curious.

It draws you in, with its welcoming pale ales and enveloping stouts. It offers you a point of difference, thanks to the stories of the men and women who wake and work at all hours. It gives back what you put in, as you learn about the styles, follow the breweries, and gain a greater appreciation of what it all means. And what it all means is this: Craft Beer is for The Geeks.

But that's OK. There's nothing wrong with being a Geek. Far from it.

Back in the day, geeks (with a small g) were socially inept, clueless types on TV, orbiting at the periphery of their respective scenes, failing to fit in. Typically, they clung to one thing they knew well – science, math, trivia – and it became a crutch they both leaned on and were beaten with. These brainiac outcasts were mocked and pitied for their lack of assimilation. But you know what? Geekery has grown up.

The tipping point was a moment of realization in popular culture. Being a Geek (with a big G) is about more than knowledge. It's about enthusiasm. Passion. If you take an interest in something and gain a desire to learn more about it, you're a Geek. If you care about something enough to talk to others, you're a Geek. And if you travel for hours to stand in line for something, you're very definitely a Geek.

These all apply to craft beer every bit as much as technology, music, or collecting things. That line could be a video-game launch or performance from the latest beat combo*, but it could just as easily be a canned-IPA release or a taproom unveiling. Ever driven to another city for a beer festival? Packed a few *#trainbeers* for the weekend? Asked a question of the person who brewed, poured, or brought you that pint? You're a Geek. And it's great.

Geeks have done more than inherit the earth; we have changed the world. And, yes, we switched to first person there. Because at BrewDog, we are Geeks too. That's the kicker; the beauty of craft beer is that the person who brewed, poured, or brought you that pint is likely also a Geek. We exist on both sides. And that's the greatest weapon craft beer has when faced with the industrial, mass-produced beers out there. Our community. Our passion.

Small, independent, local breweries are being supported and driven forward by people who realize they are a point of difference like never before. Hell, those same breweries are founded by people who realize this. It's a self-fulfilling prophecy that the big boys can never hope to understand. They rely instead on uniformity. Beers you wouldn't recognize if they were in an unbranded glass. Fake stories. Drink the beer and conform to an identity you see in an ad.

***not a music guy.**

Totalitarianism won't play here. Craft beer is for The Geeks, but Geeks don't like to be pigeonholed. It's one reason we have reappropriated the very term itself. To conform to our own identity, not someone else's.

Here's why... Here's what craft beer brings to the party and why all Geeks should rejoice...

VARIETY

From bitter, dry, and quenching craft-brewed Pilsners to earth-moving barrel-aged stouts, craft beer has it all. There's nothing else you can put in your glass that has as many flavors, aromas, or layers. We have sharp, puckering sours; funky beers made joyous by wild, windborne yeast; hop-charged pale ales that bring a smile to the face. Craft beer is the broadest of churches, here to welcome each and every Geek, every week.

RESPONSIBILITY

Reduced to a bumper sticker, it would be this: *Craft Brewers Do It the Right Way.* We proudly give back and embrace provenance, seasonality, and locality. We stand up for the little guy, and each other. Anyone can get on board with this, and the driver that is independence of thought and mind is (or should be) a principle that no self-respecting Geek should leave the house without.

BACKSTORY

Every Geek loves something to get their teeth into, and the joy of craft beer is that it is a scene that shares its stories. Embrace the power of social media to discover the philosophy of your new favorite brewery. Ask them why they brew what they do. Every brewer unshutters their brewhouse in the morning for a different reason; there are as many as the people who turn the keys. And most of us are open enough to engage on what they are.

OUTLOOK

Craft beer respects history, but not boundaries or borders. You can be anywhere in the world and discover it. This is what our scene has given to the world – the chance to find that point of difference. No longer are geeks to be found poring over old magazines or holed up in dusty museums. Geeks can enjoy an entire world of different beers whether they live in Patagonia or Pittsburgh.

HYPE

Not to be underestimated is the power craft beer has to create a movement. Individual geeks have connected into communities tens-of-thousands strong, with local chapters around the globe. Doing this lifts the tide that floats all boats and makes those who brew to satisfy balance sheets aware there's another way of doing things. A way where people with passion have the final say.

Over the course of this book, we'll break down these concepts and more, and let you discover (or revel in) your Geekery. In Chapter One, we begin with the Point of Difference, and lead with the key battleground that keeps brewers up at night – Beer Quality – dialing-in to why it matters and detailing the importance of freshness. We also take a deeper dive into the Fantastic Four and a look at how craft has changed the game when it comes to water, malt, hops, and yeast.

Next up, Culture Club focuses on the single biggest difference between craft beer and the wannabes – the culture of those who brew and enjoy it. We discuss why craft beer is for the people instead of the elite by profiling the pioneers who truly made – and are continuing to make – a difference (people and beers both). These are our heroes.

In Chapter 3: The Style Council, we chart the journey beer styles have taken as craft brewers have got their mojo going – experimentation and the lack of boundaries has created so many amazing beers. We dissect a dozen ways in which traditional styles have evolved into beers to start a movement, as craft beer changes the very style of beer itself.

We also revisit mealtime majesty in Dog Eat Dog 2.0, as we look at the science behind why beer and food work so well together. Rather than us simply telling you, we also include 18 recipes you can re-create at home, divided into six multi-course, themed pairing dinners. If there's anybody you know who remains skeptical about the nature of your Geekery, invite them over to make them see the light. We'll tell you exactly how.

Finally, we raise the bar for home-brew once again. We have another selection of BrewDog recipes and those from guest breweries to bastardize at home, many of which have never been published or even revealed before. And we look at the technical aspects of nailing the perfect brew at home; how to avoid off-flavors, brews going bad; and revel in the high-level tricks a home-brewer can deploy in their arsenal.

We invite you to join our world, where every beer is different, whether you are looking for that or until you find one you love like no other. We already proved that Craft Beer is for the People. But Geeks are People too. Beer people.

THE POINT OF DIFFERENCE

QUALITY BEGINS (AND ENDS) AT HOME

NOBODY KNOWS YOUR OWN BEER BETTER THAN YOU

Here's a job offer for you. Breweries like ours actually pay some of our staff to sit in a room and drink beer. Is this the greatest career in the world? Maybe. Although it comes with a weighty responsibility. Our sensory panelists aren't technically paid to drink; they are paid to taste (although drinking as they do so is par for the course). The reason why we pay our guys to drink our own beer is simple – they know what it should taste like, better than anyone else on earth.

Brewers work around the clock to ensure that their part of the process is as good as it can be, and the beer is exactly as they intended – and the brewery as a company then needs to follow suit to check that their work has not gone to waste, and nothing untoward has happened along the way. Every beer that leaves us needs to be amazing. And when it comes to beer quality, we are our own biggest critics.

So, what do we look for?

First up, a crucial point to make is that the previous paragraph describes a picture far more cut-and-dried than it really is. Our Quality Team doesn't just sit waiting for the finished batch to land in their hands before they test it. We delve into every aspect of a beer long before the brewhouse signs it off – starting even before the brewers do, in fact. As they slice open the sacks of malt and begin to mash in, our Quality Team are checking everything is as it should be.

Our Quality Assurance (QA) program can be split into three sections – only one of which is about the beer. We also fastidiously check the raw materials as they come in and before they are deployed. Here's how we break down the Fantastic Four to ensure they will make a Fantastic Beer:

Malt. Our QA team runs color matching to check each batch is on spec, as well as running Congress Mash Testing – essentially a miniature brewday, complete with milling, mashing, and lautering. The congress wort is checked for gravity, pH, color, and any number of other things that mimic exactly what the brewers will then be discovering as they use the remainder of the batch for their big brew. Our QA team finds it all out first. Forewarned is forearmed.

Hops. In our lab, among many other quietly whirring pieces of equipment, is a gas chromatograph that we use to analyze every shipment of hops we receive. This technique separates the compounds within the hop oils and key volatiles, and analyzes them in a way we can compare with the data received from the hop supplier. Making sure that everything matches means the hops are good to go.

Water. We have a massive water-testing program – we run a full ionics profile of our mains and brewing water, with the latter being subjected to ultra-filtration and UV treatment. Every water source in our brewery is tested every week of the year for hardness, pH, and a range of dissolved minerals such as phosphate, sulphate, sulphite, and iron. Plus, we pick up any bacteria or wild yeasts floating around (hence the UV treatment).

Yeast. Speaking of which, as our yeasts have the most important role to play in our brewery, and as they are a colony of living organisms, you can bet your bottom dollar that we analyze everything we can about them to make sure they are happy and healthy. Our team of microbiologists gets up close and personal with our different strains before, during, and after they are pitched into the business end of the brew. A successful brewery owes everything to its yeast.

Aside from the raw materials, we have to keep on top of every packing line, all of the cleaning systems and pipework (including all of the water used there as well), and all kinds of other parts of our brewery that fit under environmental checks. Our brewery is certified for Food Safety and Health & Safety – which our crew works incredibly hard to achieve – and demonstrates that QA is about much more than whether the beer tastes right or not. On that front, we run a whole suite of tests, checks, and data-collection points on the beer (always, by the way, or it never leaves our brewery). Deep breath…

Full Microbiology – checks for yeast, wild yeast, and any beer-spoiling organisms that shouldn't be there.

Analytics – gravity, ABV, pH, IBU, color, haze, VDK (vicinal diketones, a.k.a. diacetyl), low-boiling-point volatiles (for example, acetaldehyde, DMS, various esters, and hop volatiles). In short, any off-flavor that can be uncovered via chemical analysis.

Packaging Quality – levels of O2, CO2, N2, over- and under-fills, package quality, label quality, can-seam integrity.

Sensory Analysis – quantitative descriptive analysis for new beers, true-to-type tests for existing beers.

IS THIS THE GREATEST CAREER IN THE WORLD?

Our lab guys are always busy. But we need them to be – as a brewery we live or die based on the contents of your glass, and if a beer has the sweetcorn whiff of DMS or if (heaven forbid) our brewing system picks up an infection of wild yeast, we need to know long before you ever would. We even check things like whether our bottles are too full, or not full enough (there's a philosophical joke in there somewhere).

Our amazing lab crew – and those who do similar jobs for craft breweries around the world – are the last line of defence each of us has. Without their commitment and skilled eyes, noses, and tastebuds, more of the beer you buy would end up going down the sink after a single sniff. Our Quality guys have our backs every day of the week. Sometimes, getting paid to drink beer has even more advantages than it would at first seem …

THE FANTASTIC FOUR 2.0

Beer moves on; nothing remains static. However, as we push forward into this brave new world of multiple flavors and constant releases, one thing is truer now than ever: at its heart, beer is about four things. As the new powers of social media, beer ratings, label art, launch and the rest take craft beer into new areas, without those four things you wouldn't have beer at all.

Water. Hops. Yeast. Malt. Still the fantastic four.

Every home brewer begins their adventures with these four things and whatever rudimentary kit they can scrape together (or, if they are lucky, have bought for them). Every commercial brewer begins their brewday with these four things on their large-scale kit (that, if they are lucky, has been bought for them). These four things are the foundation of beer and the structure our entire industry is built on.

At their simplest, they are used to bring out the best in the others. Without malt, yeast would have nothing to feast on. Without hops, the resultant liquor would be insipid and lacking flavor, bitterness, or aroma. And without water – well, that would be a bowl of disheartened yeast slowly glooping over a pile of cereal and plant matter. Beer needs all four to come together – as well as the brewer, always the fifth wheel.

In truth, though we say "the brewer" a fair bit in these pages, what we really mean are the dozens/hundreds of people at the brewery. Many craft brewers are one-man or one-woman bands, which is amazing, given all the roles to perform. Our shorthand for everyone at the brewery maybe leaves out 99 percent of the people who have an input into what makes a great beer.

Back to basics: water, malt, hops, yeast

However many hard-working people there are, they owe everything to these four ingredients. They are the building blocks but also incredibly complicated ingredients, each with a backstory. Behind every sack of grain that arrives at a brewery there are those at the maltings, and the farmers before them. Hops likewise. Water has a myriad of tales. And yeast – well, yeast is just bonkers.

Let's dig into each a little more and see where it takes us.

FANTASTIC FOUR

HOPS

Every year, like clockwork, the start of September heralds the onset of a unique migration. From all parts of the globe they come, navigating many different routes leading toward the same destination. Those that arrive first are few,but over time this transient population swells to number many dozen, all returning to a land rich in a natural bounty there for the taking. Herds of brewers are on the move to renew their lifeblood. Hops.

As summer turns to fall, brewers from around the world journey to the Pacific Northwest, some of them having traveled for thousands of miles. All are searching for the same thing – the best of the hops that are grown in the region. Tiny starting pistols fire in unison, causing brewers to drop everything and trek from their brewery to the hop fields that carpet the mountain-ringed valleys of the top left corner of the country.

In reality, it's less the bang of a starting pistol than the chime of an inbox, as an email arrives from a hop supplier inviting them to the selection that takes place as the centerpoint of every annual harvest. Of all the many hop-growing regions of the world, one of the most important is centered on the high peaks and low plains of Washington State, around the city of Yakima.

The Yakima Valley produces three-quarters of all the hops grown in the United States. This region, around 150 miles (240 kilometers) southeast of Seattle, is also known for fruit growing and vineyards, but with nearby topographical features named Simcoe, Ahtanum, and Cascade you need only look at a map to realize you're in hop country. Volcanic soil, clear rivers, and over 200 days of sunshine a year will do that to a region.

The hop farms that dot the valley are hives of industry for only a short few weeks a year. Their hop-picker machines sit idle for 11 months before shuddering into life as August tips into September. For four weeks they work 24/7 as an army of workers slice the bines at the ground with a bottom-cutter harvester and then again from above with a top-cutter, draping the hop-laden branches onto trucks and driving them to the waiting hop trellises.

Many farms major in one or two varietals – some will grow mostly Amarillo, others Citra or Summit. Some farms grow a few; a smaller number grow many different types, with fields of experimental hops. All of them grow other things too – apples, mint, grapes. Everywhere you look there are heavily laden trees or towering fields of corn. Also everywhere you look are brewers, here to choose the hops that will flavor your beer for the year ahead.

The Morrier Ranch hops processing facility in Yakima, Washington.

Hops are available online, of course, but there's no substitute for being there and taking part in the selection. Brewers know their beers better than anyone – so why put the next 12 months' Punk IPA or Sierra Nevada Pale Ale on the roulette wheel of online descriptors? You go to Yakima and get your hands on the hops – literally – that match the aroma and flavor profile of your flagship beers (and any other beers you want to produce).

This is just part of the attention to detail that sets craft beer apart. We go to Yakima to trace our hops back to the farms where they come from – and to meet the men and women working all hours to grow its cornerstone ingredient. Brewers are control freaks. We like continuity. Hand-selecting the batches of hops that will be shipped to our cold store is the best way to guarantee that our beer tastes the way it should.

Hop selection is pure behind-the-curtain stuff. Brewers, given allotted times to arrive at a hop supplier, are met with a series of samples given over (typically) as cylindrical plugs of compressed hops. They are broken apart, rubbed between the palms, and the aromas inhaled. And that's it. That's how many of the beers you drink will be sparked into life; brewers silently bent over a lab table, grinding palmfuls of hops together and breathing in their sticky oils, before writing down thoughts, scoring them, and moving on to the next one.

Here's how it typically plays out:

HOP SAMPLES

(for example, Cascade, Tomahawk, Citra)

19-WA-241-032 (BIN D-021): Lemon. Citrus. Top Level. Lime Zest. Stonefruit. **3**

19-WA-477-008 (BIN I-003: Sharp Citrus Cheesecake. Bright. Juicy. Sherbet. **4**

19-WA-475-022 (BIN C-031): Big OG. Pine. Chives. Tomato Plant. Unripe fruit. **2**

The first thing that brewers have to do is inhale the aromas, not the actual hops (although everyone at the table has done that and doubled-up in a coughing fit that leads to looks of sympathy). Drawing in breath through a cupped handful of hops, they have to immediately critique the aromas they are getting. Quickfire notes are the only way to do it – talking out loud would influence others making their own determinations, so you reach for a pen with your sticky green fingers and write down aroma notes as they come to you.

Many craft brewers today ace the hops that give zesty, citrus highs, as these alpha acids will hopefully yield the same notes in the pale ale they are planning (or, of course, the pale ale they've been brewing for 30 years). Many shy away from "OG," as onion/garlic notes aren't what they are after. But other breweries know that this translates to dank, sweaty, resinous IPAs. And if you want New England IPA? Stonefruit is your friend, friend.

The table may hold a dozen of the same type of hop, so taking a quick break and "resetting" your most important sense by smelling the back of your arm is important. Particularly when you narrow things down to the few that might make the grade. This is when the deliberate and personal becomes collaborative. Your notes are pooled, discussions held, and a consensus reached. The hop suppliers add in their detailed analytical information, having held it back until this point so as not to sway your judgments. And then, brewers say the words their hosts have been waiting for: "We'll take these ones."

That short sentence can result in 50,000 lbs (23,000 kg) of hops changing hands at a stroke. Batches can be supplied as they are or blended together to create a precise mix of exactly the profile you traveled to the Pacific Northwest for. The table is swept, and a quick drink of water and blow of the nose is all the time you have before the hop supplier looks at you all and says, "OK, next up we have Centennial with eleven samples..."

FANTASTIC FOUR

MALT

"The art of creating good malt out of bad barley has not yet been discovered." So reads a quote, dating from 1888, that is inscribed on the wall of the wood-paneled reception of Tweed Valley Maltings. Owned and operated by five generations of the Simpson family, the maltings lies in the middle of a fist-shaped bend in the river before it meets the North Sea at Berwick-upon-Tweed, the northernmost town in England.

The quote, attributed to 19th-century brewer Robert Free, cuts to the chase – the skill, experience, and dedication of the maltster is all for nothing if what they have to work with isn't good enough. It sums up the almost symbiotic relationship between farmers and maltsters, even if it is a relationship built on commerce – the farmer sells to the maltster, who then refines the product to fulfill their contracts to brewers (and also whiskey distillers).

This is, as we have just learned, exactly the same set of tiered relationships we see with hops – farmers grow the product, sell to those who do the processing, and then the goods are made available to the brewers, with stores retained in case of shortage caused by a year with failed harvests. (Simpsons buy over a year's-worth of barley at harvest, to guarantee brewers won't be caught short when they need malt the most.)

Malt is packaged and ready for delivery at Simpsons Maltings.

Despite being the backbone to beer, malt hasn't had it easy. As industrial brewing took hold in the UK in the 1970s, the amounts ordered fell and the varieties ordered by the large brewing concerns narrowed to a few basic pale-lager malts. They were tough times for independent maltings – homogeneity is never good for those on the receiving end of it. However, the maltsters have had a revival over recent decades, from two sources.

First, the Scottish whiskey producers. Something the Yakima hop farms don't have going for them is a second outlet – and the shared starting point of beer and whiskey is a boon to those who make malt. One has always supplied the other, of course, but the interest in "Scotch" from the 1980s onwards gave maltsters a new outlook in direct opposition to the one imposed on them by the multi national lager concerns of the same time.

Recently, a second new type of customer has emerged to go in to bat for the independent malt producers – craft brewers. Their interest in the entire chain of production and their reliance on zeroing in on individual attributes of certain batches is music to the ears of those who have worked for generations to ensure exactly this, only to be pressured into simply delivering a commodity. Whiskey and craft beer have given malt a new lease on life.

Admittedly, malt is behind hops in the popularity contest of the moment. Stories of a 1-lb (450-g) sachet of rare, ultra-popular hops costing more than a 50-lb (23-kg) bag of malt are not uncommon, but the fact of the matter is that whether malted barley is overlooked or not, you can't create beer without it. The scale of the operations involved is truly colossal. Those at the cutting edge, like Simpsons, have thousands of farms around the UK that feed into them, and just as many accounts that are waiting for a delivery so that they can mash in.

"When the right crop goes into the right field, anything's possible," says Richard Simpson, fifth-generation family member and Vice Chairman of the company founded by James Parker Simpson in 1862. Like other independent producers of malt, they have worked tirelessly to refine the process over the years, with their facility at Berwick-upon-Tweed* accounting for over 240,000 tons of malt a year, which is dispatched to brewers around the world.

* Berwick takes its name from the Anglo-Saxon words for barley, "bere", and town, "wick".

As distilling and craft-brewing have increased demand, maltsters have responded in kind. They now collect data on individual fields worked by the farmers they buy from – at Simpsons these "field inputs" are a requirement for doing business, and are supplied along with the newly harvested two-row barley from each farm. A detailed set of measurements covering soil, fertilizer use, yield, and other factors, they are the secret weapons of the maltster.

"From these, we know exactly that this agronomy advice will work here, but not here," continues Richard, arms waving to indicate imaginary plots of land. Everything is accounted for and then fed into their Quality Team to be modeled. It's a win-win for both halves of the deal – the maltings receives the best quality wet barley and the farmer receives fewer rejections, where the grain is deemed not worthy of being malted and is re-routed for animal feed.

When it comes to making malt from barley, two things are key at every moment: temperature and moisture. Both are monitored around the clock at the maltings, because if either gets thrown off-kilter, the barley grains will fail to germinate, or the malt will be ruined by mold. Water content is critical – a typical grain of barley harvested and trucked to a maltings can contain around 18% water; the very first thing the maltster needs to do is reduce this to 12% through drying. And that's before the malting pathway even begins.

Once the barley is ready to go, it is steeped in water to begin the process of germination that was prevented by it being dried out in this way. This can take up to two days, with repeated soaking and drying raising the moisture content of the barley back up from 12%, to 33%, to 47% – the barley being persuaded to sprout tiny rootlets thanks to immersion in water. At Simpsons, this is then moved to a Germination Kilning Vessel (GKV), where it is germinated for four to five days while being kept moist and "turned" to break up the rootlets that bind together as they grow. This is where the proteins are broken into carbohydrates, which the brewer will further convert in the mash tun.

After germination, the barley is kilned. Drying fixes that precise moment of sprouting, and passes the sugars on to brewers instead of retaining them inside the seed to continue growing. For centuries, malt was turned by hand after being laid out on the floor. This labor-intensive process was transformed in the 1950s by the Saladin Box – an invention where drying barley was turned mechanically three or more times a day. The result was revelatory. Where previously Simpsons had 17 malting floors, they could then do everything in one place. The GKV is the modern interpretation of the Saladin Box (which sounds like something a magician would use), and is where the barley is germinated and then kilned into malt. Both acts take place in the same colossal, domed, steel chamber; huge screws turn the wet barley to unmat clumps of tangled roots, and the same screws turn the barley again as it dries in the kiln for 24 hours.

The malt finishes its journey by thundering along conveyors that can move 80 tons per hour to 650-ton silos, ready for dispatch. Of course, if we are talking about a "journey", the malt has only just started. Where the brewer takes it from here is up to them…

WHISKEY AND CRAFT BEER HAVE GIVEN MALT A NEW LEASE ON LIFE.

FANTASTIC FOUR

WATER

The Earth: 71% covered by water.
Your body: up to 61% water.
That beer in your hand? 95% water.

The craft beer industry revolves around hops, under appreciates malt, and owes everything to yeast. But none of these are beer's most important raw material. Without water, our planet, our bodies, and—most importantly—your beer wouldn't exist at all.

We're only scratching the surface with that paragraph, or "bending the surface tension" in a more fitting metaphor. Sure, up to 95 percent of beer is water, but breweries use far more water than that – they use it for cleaning, cooling, generating steam, and during packaging. The ratio of water used to water that makes it into beer is different at every brewery, but it can range from 5:1 to 10:1. In these days of drought and climate uncertainty, correct water use is even more critical than ever.

Take California. According to the National Drought Mitigation Center at the University of Nebraska, the most populous state in the country experienced some form of drought for 376 consecutive weeks leading up to 2019. That's *seven years*. December 20, 2011, to March 14, 2019, at least some part of the third-largest state had a chronic shortage of water.

Another fact to ponder is that five months later, in August 2019, California became the first state to hit 1,000 breweries. With the pioneers at Anchor, New Albion, Sierra Nevada, and the rest in the rear-view mirror (although two of those three are still going strong today), you could argue that California was where craft beer was born and raised. The California Craft Brewers Association (CCBA) has calculated that 95 percent of residents in the state live within ten miles of a brewery. If you live in San Diego, you likely live within a mile of ten (at least).

So, how do these two truths intersect? Well, with increasing difficulty, even as (or more likely, because) other industries use more water. Back in 2015, it was calculated that California's almond-growing industry accounted for 10 percent of the state's entire annual agricultural water use – more than the populations of Los Angeles and San Francisco use over the same period in a year, combined. Each single almond takes a gallon of water to produce.

Californian breweries have had to innovate to preserve precious water as it becomes a dwindling resource.

Brewing is far from hitting these numbers and, unless the number of breweries in California reaches 100,000, likely never will. But every impact on water use is an important one when the amounts available are decreasing, so increasingly brewers are doing their part. One option is water treatment. For breweries that have the budget, treating their wastewater leads to recycling and reuse in areas where where water from the main would otherwise be deployed.

Stone Brewing in Southern California managed to get their ratio of brewhouse:beer water usage down to 3:1, thanks to a rigorous treatment system that cost $8 million. In 2017, they supported San Diego's multiyear plan to purify the city's water with a beer made from purified reclaimed water – Stone Full Circle Pale Ale. Served at their Stone Brewing World Bistro & Gardens, the water it was brewed with was cleaner than regular tap water.

While the problems of water scarcity are increasing, the solutions have often been around for a long time. Another way in which breweries can use less water is to change processes at the lautering stage. Separating the wort from the spent grain following the mash can be done in a few ways, but typically many breweries spray water over the sticky mass from a rotating sparge arm, sluicing the water downward through the grain bed at the same rate as it is run off at the bottom. The bigger your vessel, the more water you need. Yet an alternative does exist that dates back to 1901 – the mash filter. Developed in Belgium by Phillipe Meura, this press-like arrangement is a series of filters through which the wort is pumped. Sparge water is used, but far less than during lautering – the wort is part-run, part-squeezed-out mechanically. Brewers argue that one gives a better quality wort than the other, but few can argue that mash filters use less water.

In the far Northwest, the Alaskan Brewing Company installed a mash filter in 2008 and almost immediately saw a resulting decrease in water usage (important even in a part of the world where water is not in short supply). They were the first American craft brewery to use one, reporting that in a year it resulted in them using nearly two million fewer gallons of water. However, they also discovered an additional effect that made it even more worth their while. The mash filter also improved their efficiency, resulting in malt usage decreasing by 6 percent – a significant sum for a brewery of their size operating in a city with no roads, where every raw material has to arrive and depart by barge. They also operate a spent grain boiler, turning grain from the mash into steam used for heating. As the press squeezes the malt dry, it ends up with a lower moisture content and is thus easier to heat – saving them 65,000 gallons of diesel fuel in their boiler every year.

So water use is important to quantify and keep in check for its own sake, but for many other interconnected reasons as well. When we talk about things that are interconnected, the environment around us immediately springs to mind. One way in which breweries can make a difference here is to channel the water that leaves the brewhouse to somewhere other than the nearest drain.

Some breweries are taking this to an incredible level by developing their own system of wetlands. Water leaves the brewery and runs through a series of ponds and ditches, filtering through plants and sediment until it returns to the natural water system as pure as nature can make it. Algae in the ponds work aerobically, during the summer months, removing things like nitrogen and phosphorous from the water and generating oxygen.

For an industry that uses a lot of water, knowing where it goes and being able to return some of it to where it came from is about as good as it gets. As the climate crisis continues, look for your local brewery to play their part in maintaining a supply of beer's most important ingredient.

EVERY IMPACT ON WATER USE IS AN IMPORTANT ONE.

FANTASTIC FOUR

YEAST

Stop us if you have heard this before, but yeast is incredible. If you sat down and considered every stage of the brewing process, fermentation would leap out at you clear ahead of any other. Sometimes literally. The power that yeasts have when they get going is remarkable, and the power they hold over us is absolute (and life affirming). Brewers and beer drinkers owe everything to these single-celled fungi that we can't see.

The relationship we have with yeast has spanned a complete arc, from ancient history, where their presence wasn't even known, to now, where we have sequenced its entire genome. In 1996, *saccharomyces cerevisiae* became the first eukaryote that scientists decoded in this way, teasing out all 6,294 genes from its nucleus. In yielding its genetic code *S. cerevisiae* directly helped us on the pathway that led to the sequencing of the human genome a decade later.

And the day job for that *saccharomyces cerevisiae*? Turning wort into beer.

As if its role in how we understand ourselves wasn't central enough already, the species that the scientists decoded was ale yeast. Hey, brewers had nothing to do with the decision, but we'll take it. Found naturally on the bark of trees, leaf matter, and fruit skins, there are many types of yeast that we use to ferment sugars into alcohol and carbon dioxide – but *saccharomyces cerevisiae* is the granddaddy of them all. It has strains that lead to any number of aromas and flavors.

Although we love them all equally, at the top of that particular pile has to be the strain that brought craft beer to the world: American Ale Yeast.

OK, so some forerunner beers, like Anchor Steam, were fermented out with lager yeast, and on the easterly side of the pond, there's the Burton strains and titans like Whitbread Ale yeast... and that's even without considering the single-cells behind all of the great German and Belgian beers of old that remain with us today, but let's focus our attention on American ale yeast because of a single reason – it is the power behind the pale.

American Pale Ale lies at the head of a family tree that has moved through American India Pale Ale and then about a dozen variants (as we will see in Chapter 3). To the nascent craft beer movement of the early 1980s, pale ales were easier and faster to produce than the European lager styles; unlike the slower, cool-fermenting lager yeasts, pale ales could be turned around faster thanks to a top-fermenting strain.

Typically, at first, these were English ale yeasts brought in from the old country, but in California, things were changing. The "Chico Strain", named after the hometown of Sierra Nevada Brewing Company, became one of the emergent home-grown American ale yeasts and sparked thousands of commercial pale ales and millions of home-brewed versions. Brewers in other countries simply followed suit; American ale yeast had become a force to be reckoned with.

The reasons why boil down to its character – American ale yeast powers away, attenuating well, but brings little to the party as it does so. It is one of the blankest of canvases for brewers who want their other ingredients to do the talking. Hop-forward and malt-forward styles shine thanks to this kind of yeast – and, if you've ever had an American-style pale ale, you'll know exactly why they revere their hops. Not for nothing is this strain now the "house yeast" of craft breweries the world over.

BREWERS AND BEER DRINKERS OWE EVERYTHING TO THESE SINGLE-CELLED FUNGI THAT WE CAN'T SEE.

American ale yeast allows you to produce heavily hopped beers that are very clean in profile, with no yeast-derived esters or higher alcohols present to affect the flavor of the final beer. In essence, with American ale yeast, what you put in, you get out. And with its attenuation holding even at higher gravities, it also lends itself perfectly to Double IPAs that are omnipresent in cans today. American ale yeast is the best all-rounder there is.

Of course, for every all-arounder there are many more specialists. Everything has an opposite, a shadow. Where there exists recently domesticated obedience, there must also be an untameable wild child. The yin of clean must have the yang of... dirty.

The blazing light and noble intentions of American ale yeast have an opposite number: the British fungus.

As craft brewers have experimented with differing flavor profiles and stretch the limits of unusual styles, they have reignited the thrill of alternative fermentation. Leading that particular craven horde into battle against the piously clean American ale yeast is the magic, the mystery, and the (sometimes) malevolence of brettanomyces. Like its cousin saccharomyces, it too has gone full circle.

The evidence of its existence was first described in 1904, after Danish scientist N. Hjelte Claussen cultured a variety from English stock ale – a long-gone pale ale/bitter style brewed at high gravity and stored in barrels for up to a year. This yeast was eventually named after him (*brettanomyces claussenii*) and Claussen and his beer-drinking contemporaries saw its character as a favorable one – probably why he was interested it culturing it in the first place.

However, brettanomyces is a tricky customer. With high levels of phenols and esters, it produces flavors that are decidedly at the strange end of the spectrum, and its characteristic tang fell out of favor. Over time, these quirks of flavor became faults of production and brettanomyces became the enemy. Clean, crisp lagers and ales took charge and the wild, unpredictable, aged beers went the way of the dinosaurs. The weirdness takes some quantifying:

Leather. Horse-Blanket. White Pepper. Funk. Earthiness. Band-Aid. Tobacco. Barnyard. Wet dog.

These are all genuine flavor descriptors recorded by people trying to sum up what beers fermented with brettanomyces taste like. Accidental inoculation following some kind of infection leads these flavors to be harsh, as the yeast isn't in the zone it likes to be in. If you intentionally use brettanomyces and give it a friendly environment, such as a high fermentation temperature and low hydrostatic pressure, the flavors will be peppery, with fruity, tropical esters.

Two things: first, you may now realize why beers with these wild yeasts meant the use of brettanomyces faded from the limelight. Second, by now it won't surprise you that the challenge of deploying it correctly to yield the best results is right up craft beer's alley. Reborn, retooled, and rebranded as simply 'Brett,' wild yeasts are back in fashion. The circle is complete.

As craft beers are intentionally inoculated with Brett and aged in wood, again they have found a new audience. The strains are in vogue – those that we know about, anyway – there are likely to be many more hiding in the grain of old barrels we will never uncover (brettanomyces is so insatiable it ferments every sugar source in the wort and then can break down the sugars present in wood). Brett beers are a unique throwback; a liquid timeline that embraces aging, blending, and vigor. They also result in wacky adjectives on beer labels – but then, that's craft beer in a nutshell.

BEASTS FROM THE YEAST

THE BEST OF BRETT & AMERICAN ALE

As we have seen, these twin pillars of modern fermentation have very different backstories. American ale yeast is the original kingmaker of craft and brettanomyces is the new kid on the block, returning triumphant. Both have huge influence and are modern players steeped in history. As ever, the best way to discover their quirks is to flip off the cap of a bottle, or crack open a can, and pour into your nearest glass.

Here are a few pivotal beers that are powered by the might of American Ale and Brett:

AMERICAN ALE YEAST

SIERRA NEVADA PALE ALE

Origin: Chico, California

ABV: 5.6%

IBU: 38

Key ingredient: Cascade hops

It's the Cascade hops that take the plaudits – deservedly, as Sierra Nevada pioneered their use in the US (although see page 60 for a transatlantic take), but it's the "Chico Strain" that is the secret to letting these hop flavors come to the fore. Home-brewers can pick it up commercially as Wyeast 1056 American Ale™ or White Labs WLP001 California Ale Yeast®, but Sierra Nevada have it on site, whenever they need it. Clean, with a slight hint of estery orange in the background, this pale ale is a masterclass of brewing and fermentation.

TRÖEGS HOPBACK AMBER ALE

Origin: Hershey, Pennsylvania

ABV: 6.0%

IBU: 55

Key ingredient: Crystal, Nugget hops

When you deploy a yeast that allows the hops to shine, it helps if you have a hop-back. At Tröegs, theirs is a 12-foot (3.7m) high vessel where hot wort meets and greets whole cone hops so that the two can swirl together and get better acquainted. The flavors of the latter infuse the former and create beers like this incredible amber ale, resinous and zesty on one hand and toffee-like on the other, all underpinned by the canvas of ale yeast.

FALLEN BREWING GRAPEVINE

Origin: Stirlingshire, Scotland

ABV: 5.4%

IBU: 43

Key ingredient: Taiheke hops

Right in the middle of Scotland (give or take) is the town of Kippen, which 50 years ago was famous for an enormous grapevine, said to be the largest in Europe – which is something considering Scottish wine isn't exactly a market leader. More than 300 feet (90m) across, it was cut down in 1964, but lives on in the name of Fallen Brewing's New World pale ale. The big, oily floral citrus from New Zealand's Taiheke hops forms part of the hop bill that US-05 American ale yeast gives center stage to.

BRETTANOMYCES

ORVAL

Origin: Orval, Belgium

ABV: varies (but typically around 6%)

IBU: 32

Brettanomyces sees your American Sierra Nevada Pale Ale and raises you Orval. First brewed in 1931, this should be in the top five on any beer list. The only beer released to the public by the brasserie of the same name overseen by the Trappist Cistercian monks of Orval Abbey, this is the real deal. The beer is bottle-conditioned using brettanomyces (a strain of *B. bruxellensis*) and over time the flavor changes enormously. Taste a new bottle alongside one six months old for the best indication of what Brett can do to a beer as it ages. Orval is, in every way, unique.

CROOKED STAVE ST. BRETTA

Origin: Denver, Colorado

ABV: 5.2%

IBU: 25

Denver's Crooked Stave was founded by Chad Yakobson, also known as the "brettanomyces Guru." He wrote a thesis on Brett and then released it as an open-source document for the greater good. So you know their beers are going to be able to use its power like few others. St. Bretta is a saison brewed with citrus, lemongrass, and coriander and tastes every bit as amazing as it sounds–what else from a brewery that embraces Brett as a focal point for their entire being.

CANTILLON KRIEK

Origin: Brussels, Belgium

ABV: 5.5%

IBU: 25

The iconic backstreet Brussels producer Cantillon have harnessed brettanomyces by deploying it alongside souring bacteria, and gained mythical status as a result. Unlike Orval and St. Bretta, this Kriek has the sharpness of acetobacter alongside the Brett acting as a perfect mirror to the sharp and sweet of the cherries. Marrying a wild yeast with a souring bacteria to bring out the best in a fruit renowned for a dual-pronged flavor is a stroke of mastery. All of Cantillon's beers are incredible, but this most-famous of the fruit beers is the stuff of which legends are made.

BEING SPONTANEOUS

THE POWER OF WILD YEAST

Breweries spend untold amounts of money, time, and perspiration ensuring their colonies of yeast are ready to go when they are needed. Culturing up these tiny flurrying worlds of life and giving them exactly what they want – food – so that we can all enjoy the reward is essentially farming. These microscopic creatures are nurtured by brewers as the most important part of the entire process. Yet, as with everything else, there exists an alternative. Where cultivated, domesticated yeast is tended by brewers, there are wild, untamed yeasts out there ready to eat for free.

In truth, yeast is almost literally everywhere – you're covered in it right now. Yeasts exist on the skin of fruits, the petals of flowers – even on your eyelashes. Microbes are all around (and, of course, inside) us. So it stands to reason that, for every carefully guarded colony in a brewhouse vessel, there will be many, many more individual cells wanting a piece of that sweet wort action.

As we have seen, breweries run quality checks to keep these yeasts out – until they are actually wanted. There exists a throwback way of brewing so out-there that, in order to create the beer, the brewers have to leave a set of windows open at night. They are fishing for the smallest fry imaginable – wild yeasts, floating on the breeze – and before they have the chance to land on anyone's eyelashes, they plop in their untold numbers into a warm, sugar-rich bath of gently steaming wort.

To harness the power of spontaneous fermentation, brewers pump wort into coolships (or coolers, as they are sometimes called in the UK), which are essentially giant copper baths, wider than they are deep. On dry nights in the winter months (generally October–March), the brewers throw open the windows and allow whatever the wind brings to settle in the coolship and start the magic of fermentation. It's about as incredible as brewing gets.

Actually, it is very much as incredible as brewing gets. To give the microbes the best chance of success, the wort used for their moonlit dip is a special one, produced from an intensive mashing program to create a full turbid mash that is very high in unconverted starch, maximizing the food for the wild yeast and bacteria. The initial malt bill is heavy with unmodified wheat (as much as 40 percent) and is very thick. A series of temperature rests then take place, where portions of the wort are removed, heated, and reintroduced (or hot water is added instead).

This mashing process is tough, labor-intensive, and results in a mash that is almost milky-white in color – as opposed to traditional mash run-off, which should be as clear as possible. But that's what the wild yeasts need to get going – a massive source of food, and the right type of it. The starchy wort is full of carbohydrates that cannot be fermented by cultures of brewer's yeast, but are fermentable to wild yeasts such as brettanomyces.

The next morning, the brewer arrives back at work with clean eyelashes and a spring in their step. The cool wort – which is by now teeming with well-fed wild yeast – is moved into a tank and gently circulated. This ensures an even mix of the yeast throughout, as until this point most of it is floating on the surface of the wort.* It also gets the wort ready to be transferred again into oak barrels, where the yeast and bacteria get down to business for as long as they are able to. The brewer needs to ensure each has a "blow-off tube" leading from it, as they go to town on the big supply of sugars, resulting in rows of casks that are literally foaming and frothing away.

*** Wouldn't you?**

If this entire process sounds archaic, it's because it is. The spontaneous fermentation process is used primarily for lambic, which is either sold as is or blended into gueuze. These are formative beers, fascinating to drink (and to watch others drink) – and all of their unique flavor arrives from how the yeast meets the wort in the first place. All that takes is the lifting of a window latch and remembering to turn out the lights.

Here are a few pivotal beers that are powered by the night flight of wild yeast:

IF THIS PROCESS SOUNDS ARCHAIC, IT'S BECAUSE IT IS.

CANTILLON GUEUZE

Origin: Brussels, Belgium

ABV: 5.0%

It says so on the label – 100% Lambic Bio – underneath the unmistakable image of the *Manneken Pis*. From their Brussels-backstreet location that is a living museum, Jean-Pierre van Roy and his team create spontaneous beers that are prized around the world. Cantillon are proof that brewing is an art form, and their gueuze – which starts with at least ten different lambics being selected and sampled to determine what half-dozen form this masterpiece – is proof of that.

ALLAGASH COOLSHIP PÊCHE

Origin: Maine

ABV: 7.3%

The wild winds of Maine bring the yeast to the party taking place in the coolship at Allagash. So much of a good time is had by the things that land there, the guys at the brewery have nicknamed the vessel where the post-coolship wort is circulated the "horny tank". This beer then gets a long time to cool its jets – it ferments for two years on oak with the final five months joined by farm-fresh Maine peaches.

BURNING SKY COOLSHIP RELEASE

Origin: East Sussex, England

ABV: varies

Mark Tranter has pushed the boundaries of brewing in the UK for decades, from Dark Star to his own brewery at Burning Sky. They are, quite simply, one of the top five breweries in the country and when they took delivery of a coolship and started a journey into spontaneous fermentation, a lot of people sat up and took notice. The early beers more than justify it – balanced acidity and a funky, stonefruit-skin flavor abound.

BOON GEUZE MARIAGE PARFAIT

Origin: Lembeek, Belgium

ABV: 8.0%

Frank Boon is a world authority on blending lambic into gueuze (or geuze, as they call it) and this one, as the name suggests, is a perfect marriage. Three-year-old lambic is blended with a small amount of young lambic not long from the coolship, giving a deep and refreshing sour vinousness from the oak aging but also yielding lime zest and other palate-enlivening acidity. It is nothing less than a world classic, and a must-try beer.

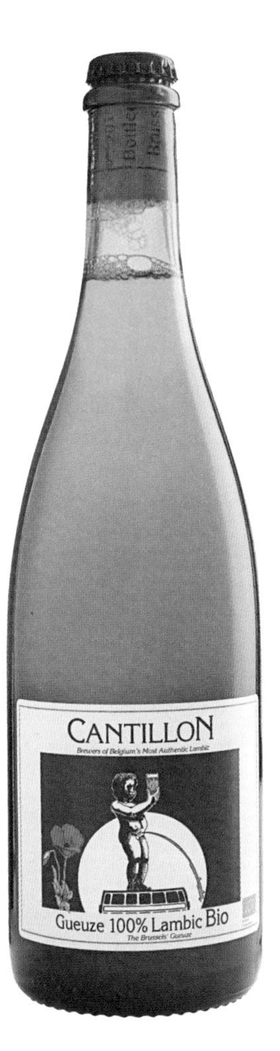

THE SCIENCE OF FLAVOR

We've established what you already knew – beer is awesome. But there are many reasons for this, including its variety, the stories of the people behind it, and how it can be re-created in your kitchen whenever you like. We'll cover all of these in due course, but let's take a quick look at another reason why hops, malt, water, and yeast make magic together. As with any group coming together – it's all thanks to Chemistry. With a big C.

Let's dig down now into the Chemistry of beer, or – more specifically – beer flavor.

During the brewing process and following its completion as the beer conditions and beyond (if the beer is then placed in wooden barrels), science is doing its best to reward you with as unique a beverage as possible. Of course, this isn't exclusive to beer – other foodstuffs that are fermented and/or processed will have a similar chemical pathway (and therefore end with similar flavors).

So, what is it about the basic ingredients that make them critical in creating the flavors of beer? It is the building blocks of the ingredients themselves. Compounds such as sugars, small proteins, and enzymes react together, adding layers to the flavor profile at each production step. And the chemical processes responsible for this flavor pile-on are known (broadly) as flavor chemistry.

Take the Maillard reaction. This is a fundamental interplay between sugars and amino acids in the presence of heat. Although it sounds as though you'd need a labcoat and Bunsen burner to see it in action, all you actually need is to pad into the kitchen in your slippers and drop a slice of wholemeal into the toaster. The Maillard reaction happens every time you toast bread, pan-fry a steak, or bake a cake. And it happens turbo-style when you turn your back on the toaster.

In beer, this reaction happens when malt is roasted, causing not only differences in flavor, but also color. Obviously, the higher the temperature or longer the roasting period, the darker the malt. This will also change the flavor of the malted barley as well as the color, and both are dependent on not just the heat applied but how the maltster controls the humidity. From pale malt to coffee malt, the flavors created at this crucial stage – way before a brewer has even turned up to their brewery – determine how the final beer will taste.

Another process that creates major complexity in beer flavor is fermentation. It transforms the humdrum into the holy cow. And the key mover and shaker here is the mighty brewers' yeast. It is responsible for the production of many of the flavor compounds in beer, transforming wort sugars, and amino acids into esters, phenols, and higher alcohols (to name a few).

THE SENSITIVITY OF THE HUMAN NOSE IS HIGHER THAN ANY EQUIPMENT CAN ACHIEVE.

Esters add fruity, floral, and spicy notes to the malty, caramel-flavor backbone of the wort. Some phenols contribute a spicy, clove-like aroma to style-appropriate beer (if you love German wheat beer, then 4-vinylguaiacol is the phenol to thank). Others, however, are an indication that things have gone awry. Beer infected with wild yeasts or bacteria often have 4-ethylphenol, which causes horse-blanket/barnyard aromas. Poor choice of brewing water can yield chlorophenol and its resultant Band-Aid/medicinal aromas. No thanks.

The third part of that fermentation triumvirate is the most commonly recognized (for better or worse): alcohol. The most popularly known alcohol in beer is ethanol, a yeast by-product that makes beer an effective social lubricant. Other than ethanol, yeast also produces higher alcohols, some responsible for floral aromas (phenylethyl alcohol for aroma of roses), but mostly for warming, solvent-like aromas.

Aromas in beer during fermentation tell us a lot about how healthy or unhealthy the yeast is. Two compounds used as indicators of yeast health are diacetyl (buttery aroma) and acetaldehyde (emulsion paint, green apple aromas). Too much wort sugar causes increased acetaldehyde production and unhappy yeast cells that cannot absorb excess diacetyl and convert it to odorless 2,3-butanediol.

We rigorously monitor these aromas throughout fermentation. If yeasts aren't absorbing the excess diacetyl, it ends up in the beer and wafts upward from your glass. That ain't good. Brewers need the yeast to convert it to the odorless compound – the good news about diacetyl's obvious tang is that it enables brewers to take action and ensure the yeast has the best conditions to make desirable flavor compounds instead of negative off-flavors.

Last, but certainly not least, a major flavor boost for the beer is dry-hopping. Hops are selected for their aromas, of which there is as wide an array as there are varieties available, and in addition they are a rich source of essential oils responsible for grassy, spicy, citrusy, piney, and fruity aromas, to name but a few. Think fresh herbs added to cooked dishes: tomato pasta sauce with basil and oregano, or roast chicken with thyme and rosemary, to give you a mental picture of how hops boost flavor in beer.

Any chemical reaction can be measured, and in flavor chemistry this is relatively easy because you can smell and/or taste the effects. Having a sensory panel, with trained people to sniff out specific aromas, is a very valuable resource in a brewery.

A good, trained sensory panel will detect off-flavors in beers during production faster than any measuring equipment can. For real. This is because the sensitivity of the human nose is higher than any equipment can achieve, even if our robot overlords are closing the gap. Having said that, people become fatigued, so we double up with lab measurements of these compounds, to back them up.

To do this, one of the most common techniques we deploy in the lab is the catchily titled Gas Chromatography – Mass Spectrometry (GC–MS). This method measures volatile aroma compounds by separating aroma into its constituent aroma compounds. It pares everything back to the bare bones, allowing us to see the different aroma compound classes, mainly esters, and can be further enhanced by including a human assessor alongside the equipment.

GC–MS not only tells us the volatile compound profile, but also which of these compounds are the most important in terms of aroma creation. It also illuminates which and when aroma compounds are formed and/or degraded, and which survive the process and are carried through into the bottle and enjoyed by us. It enables us to dial in to every aroma and flavor aspect of a beer before it is finished. And yes, they don't come cheap.

Any great craft brewery should place the utmost degree of importance on flavor quality, so it is crucial that it is monitored throughout beer production. We love science. You love beer. That's maybe a more succinct way of putting it.

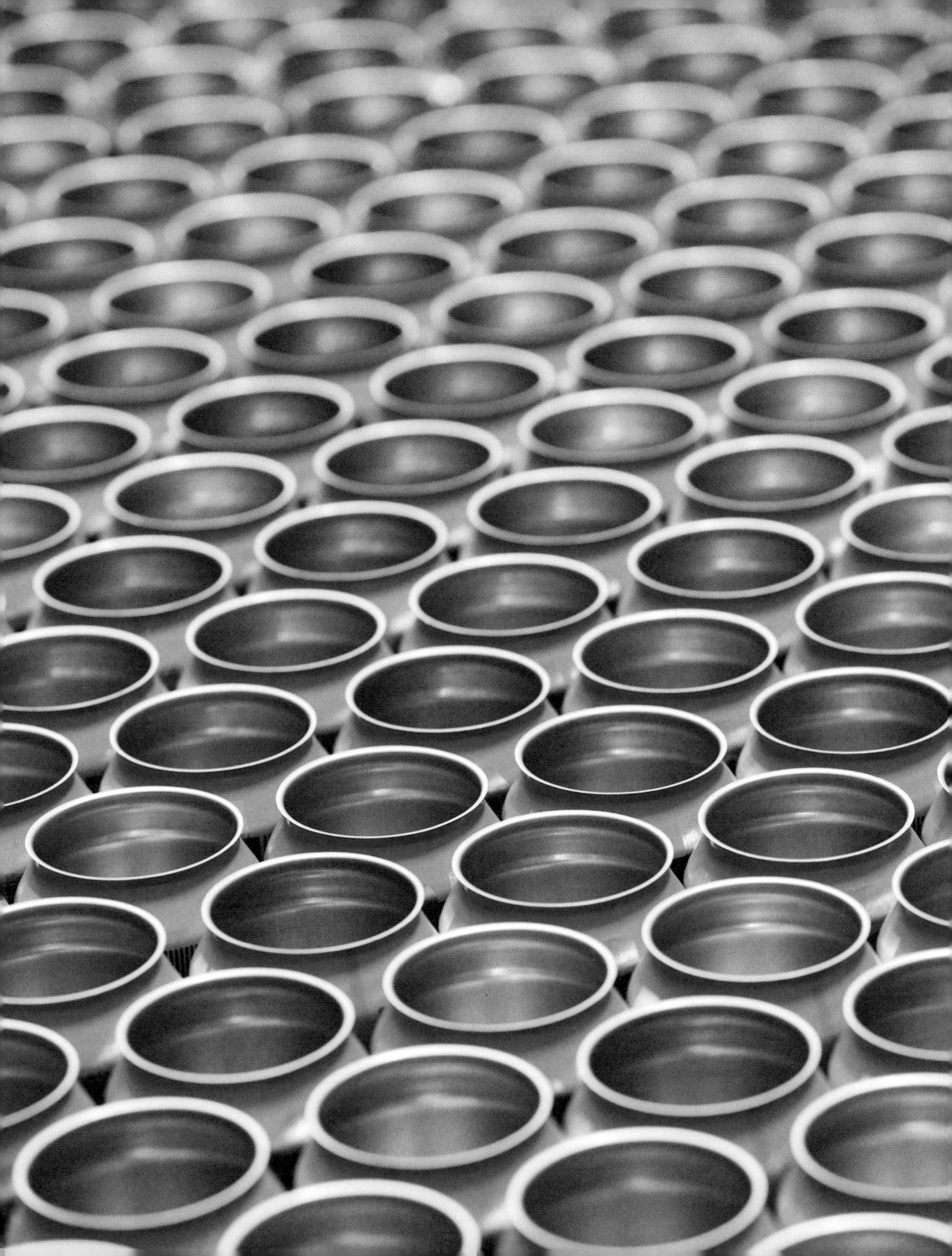

QUALITY & FRESHNESS

You never get a second chance to make a first impression. This is as true for craft beer as it is for that all-important job interview. Sounds strange, but bear with us. On that anxious morning before the big interview, you polish yourself up and practise those responses to questions you know are coming, because you are selling yourself to people who have never met you. As your paths cross for the first time, they learn what you are about, and hopefully you impress them enough to put yourself ahead of the pack.

It might be trite to say that every beer is a liquid interview for that brewery, but the analogy holds. We work in a crowd, too; if you look at just two brewing nations – the USA and the UK – at present there are nearly 10,000 breweries combined. If each of these releases only a dozen different beers a year, it gives the thirsty people of both countries 120,000 choices to make on an annual basis – and that's excluding all of the incredible beers imported from other parts of the world.

To succeed, you have to stand out from the crowd. You can do that in any number of different ways – size, spending, PR, aggression, quirkiness. But the best way? Quality.

Beer quality is the new battleground, but in truth it should be at the core of every brewery. As craft beer starts to chip away at the monopoly of the large industrial breweries, every pint matters. The thousands of small, independent craft breweries have a collective responsibility to avoid a scenario where people new to our way of thinking get a poor first experience of craft beer and never return. As a brewery, the quality of your beer should be your greatest concern.

Increasingly, a central tenet of beer quality is freshness. The time it takes to get your beer into people's hands matters. And, ironically, the greatest weapon in this new frontier has been displayed in public for decades by the big boys. If you close your eyes and think of any major lager brand's advertising, chances are that one of the images that will fill your head is that of an ice-cold beer bottle. Frosted glassware. Frozen tundra. On the one hand, serving your beer so cold that it sends jets of pain through your teeth enamel is a) undeniably refreshing, but on the other hand, b) it makes up for the lack of flavor in your product. But we also know that cold beer is better for the entirely opposite reason. If you store, transport, and cellar in cold conditions, it keeps the flavors and aromas of craft beer intact.

ALL BEER IS ON THE CLOCK.

Let's look at it this way – all beer is on the clock. It leaves the brewery tasting exactly as the brewer intended, so you'll want to get your hands on it as soon as possible in order to discover those flavors. Now, unless you live within walking distance of the brewery, that might be a tall order (this may help to explain why taprooms have surged in popularity within the last couple of decades). Chances are, then, the beer has to travel to where you can buy it. If it is chilled all the way, it will benefit everyone. It's all about the cold-chain.

Cold-chain is a supply chain where every link along the way is performed under a certain temperature. It is not new, by any means – the global food industry has been cold-chaining produce around the world for decades – but brewers have now embraced it as a method to prolong the flavors they have imparted to their beers, and make those beers stay in peak condition a bit longer or travel a bit further. It's a serious, and costly, business.

From entire warehouses maintained at a certain temperature (typically under 43°F/6°C), to climate-controlled vehicles, to breweries only supplying outlets where the beer is kept refrigerated – cold-chaining comes at a price. But the benefits of this time and expense to the beer drinker are huge. Cans and bottles left on warm store shelves for who-knows-how-long cannot compete with the same beer kept at the same temperature and given to you faster.

TO SUCCEED, YOU HAVE TO STAND OUT FROM THE CROWD.

Once the beer has been packaged, the volatile hop oils that the brewer spent time and effort getting into the beer will start to fade. Proteins from the malt will deteriorate and the beer will lose body. The bigger the flavor, the more there is to be lost. Try it for yourself; buy an IPA and leave it on your bookshelf for a couple of months, then buy another and keep it in the fridge for 24 hours. Open both at the same time and try them, side-by-side. The newer beer will have heights of aroma and flavor that the old one will not. The citrus and resinous peaks will have been dulled, their elevation ground away by the ravages of time. With beer, the ravages of time aren't measured in decades–as with our increasingly wrinkly faces – but in months. Nobody should be drinking hop-forward beer more than a couple of months after it leaves the brewery. That would be like turning up to your job interview two months late.

PUNK IPA
ELVIS JUICE

DOGTAP
DOGTAP

GET FRESH

As we've hinted before, the best way to enjoy fresh craft beer that tastes exactly as the brewer intended is to show up at your local taproom and bang on the door until it's time for opening. Once you are inside (assuming you haven't been carted away by the police), beery nirvana is but a short wait away. Yet craft beer is a global movement. Maybe your favorite brewery is in the next county, a favorite country, or another continent. What then?

Well, hopefully that favored brewer of yours has embraced cold-chain delivery and you can make it to a shop that uses chilled storage. If not, you can follow a few basic rules to give yourself the best chance of enjoying your beer, instead of wrinkling your forehead and wondering what all the fuss was about. Next time you reach for your sturdiest carrier bag (reusable, of course) and head for your beer retailer of choice, make note of the following:

HOW TO SCORE YOURSELF BREWERY-FRESH BEER WITHOUT GOING TO THE BREWERY.

* FRIDGES OVER SHELVES

If you're like us, you'll always gravitate toward the fridges if your shop has the option. If it's all fridges, then hallelujah. If it's all shelving, then never mind – the pointers below will still will help you uncover the best beer around.. Just avoid those bottles that are in the sun, if you are lucky enough to live somewhere it streams through the windows. Head to the shady shelves and work through the following rules.

* JUST ASK

This is, of course, the most obvious way to find fresh beer – go on over to the person behind the counter and ask what's just been delivered. Maybe you follow them on social media and they do a great job of advertising recent arrivals – you'll know they are among the freshest gems you can find inside. It could be that the staff are unpacking them as you walk in, so you can lift the latest offerings directly out of the brewery packaging. Can't get fresher than that!

* CHECK DATES

The rise in the use of computerized date and time stamps is a thing to celebrate for those looking for super-fresh beer. The fact that every brewery uses a different system only makes it more of a challenge. Raise those cans above your eyeline and scan the base (maybe don't upend them if they contain sediment). Check the "best before" date, if it has one. The later it is, the longer it has been around. And if the can/bottle has a "canned on"/"bottled on"/"born on" date, happy days!

* KNOW YOUR STYLES

If you have a bit of background knowledge you can quickly discover what's good throughout the year – for example, when the leaves in Europe start to turn it means German Festbier will soon arrive in time for Oktoberfest. The September hop harvest leads to the sudden appearance of wet hop beers (see page 89) – and you obviously know when pumpkin beer will turn the shelves orange. Year-round releases are good to go whenever, but obscure styles have a time and place.

* DON'T JUMP THE GUN

On the same train of thought, if you are hopping from one foot to the other at the prospect of Oktoberfest, don't let it get inside your head to the extent that you clear the shelves in July – as that's all you'll be doing; clearing the shelves so the bottle shop can get the new ones in when they arrive. As Gold Five said, "Stay on target". Eleven-month-old beer won't taste as good as another month of waiting will in order to drink the brand new stuff.

COLD-CHAINED BEER

In early 2019, we decided to take the logical step in our mission to move beer cold from brewery to glass. We designed, brewed, and then cold-chained a beer that was a celebration of freshness all the way. It was a perfect depiction of how quick-moving beer can retain its flavor and, of course, it had to be an India Pale Ale.

We put our logistics on show with a beer designed to be on the clock. The beer was 30 Day IPA, and we sold it only to those places where it was guaranteed to be stored cold end-to-end. Thirty days was its shelf life. No more.

A 30-day shelf life means a beer with a short life – but that was the point. We brewed a 6.7% ABV West Coast IPA and heavily dry-hopped it with Mosaic, Citra, and Chinook to give massive aromas of mango, yuzu, lychee, citrus, and resinous pine. Adding enormous aroma hops at the dry-hop stage can really let a beer hit the top notes. Those top notes are imparted by volatile hop oils that give incredibly fruity aromas. They are fragile, and drop away after a short period of time – as do the tropical aromas through an interaction of yeast and hops in a process called "biotransformation." If you heavily hop a beer, it's something that will happen to them all.

However, as we have seen, this natural flavor deterioration process can be slowed down by ensuring beer is stored cold, and *that* was the reasoning behind 30 Day IPA. Yet it was more than just reasoning – the solution of storing cold and drinking fresh was made real by cold-chain process: the trucks, the cellars, the warehouse, everything down the line – chilled.

To fully maximize this we concentrated on a single aspect of the logistics – the storage. Trucking things cold is easily done, if you have a refrigerated container. Serving things cold should also be fine if your cellar has a thermostat and a group of people working inside and above who know their stuff when it comes to conditioning and cellaring beer. But the storage? How do you commit to a chilled storehouse? In our case, you build one.

Just outside Glasgow, we opened a 130,000-square-foot (40,000-square-meter) warehouse that became Europe's first fully refrigerated beer warehouse for national distribution. Code-named "Hop Hub", it is outfitted with six colossal chillers that blast out cold air 24/7, ensuring the millions of cans of craft beer it can receive every day never experience ambient temperature. It became a time capsule, protecting our beer from the perils of heat damage and flavor deterioration.

It also became the key part of the chain that took 30 Day IPA, brewery-fresh, to where it needed to be (as well as all of the other beers that leave our brewhouse). Hop Hub was built with one thought in mind – beer quality – and 30 Day IPA existed (briefly) to fly the flag for freshness and quality and to change how people think about beer itself. We believe in good beer and we passionately believe in fresh beer. In fact we'd say good beer *is* fresh beer.

GOOD BEER IS FRESH BEER.

CULTURE CLUB

THE CULTURE OF CRAFT

Many people have tried to sum up craft beer in a pithy manner (ourselves included), but maybe the best two-word description to use is this: Community Rules.

Craft beer is powered by people. Beer has always been a great social lubricant, but the way in which the modern craft beer scene has brought together the people who drink it is amazing. It's amazing to see and even more so to be a part of. Brewers work their socks off and their customers have their backs. In return, brewers know that they aren't painting pictures nobody will see. Their audience is out there, and growing.

More than ever before, beer drinkers can support their favorite breweries. Whether that's by crowding out their taprooms, touring the breweries and speaking to those working there, or lining up round the block for the latest beer release. This is a feature unique to the modern age; people willing to wait in line, outside, for a few cans of beer. But the beauty is that you don't need to go this far to maintain a connection to your favorite brewery – you just need to reach into your pocket for your phone and thumb it to life.

The way companies use social media has changed almost as many times as the number of years it has been with us (even if it seems like it's been with us forever). Take it from us. Back in the day, we told people what we were up to and hoped they dug what they were reading, or watching. Today, social media is a two-way street. Our community is empowered to deliver instant feedback and to keep us honest. They can get to know us, personally, in a way that a few years ago could only be done by rapping on the shutters.

It fosters a community underpinned by relationships built on respect.

Since the enormous rolling boulder of social media flattened our free time, another phenomenon has appeared that has become the ultimate metaphor for people power and the culture of craft beer: crowdfunding.

"Support your local" now has an entirely new and improved meaning. The ability to buy a piece of a brewery – whether down the road or across the planet – has changed the game. It has given anyone the chance to back their favorite and take part-ownership. Influencing the decisions of the brewery that creates your fridge go-to is a game changer. You can literally pay it forward every time you visit your local shop.

otwarte
klatki
BREWDOG

BREWDOG
BITTER
BEERS

When we launched our crowdfunding campaign Equity for Punks back in 2009, we had no idea how it would play out. We wanted to create a community of BrewDog supporters and take them with us on our journey. We had a single caveat – the money our investors raised, every penny of it, would be put back into the company to grow it for those who came on board with us. Not fat-cat shareholders. Beer drinkers.

The response was phenomenal. Thousands of people invested in our anti-business business model, and we invited all of them up to our brewery for our first AGM (Annual General Mayhem) in December 2010. If you were in the UK then, you might remember that winter made history for another reason – epic snowfall and a low temperature recorded in Scotland of -6.3 °F (-21.3 °C). But, even still, people came. Our community was born.

CRAFT BEER IS POWERED BY PEOPLE.

Since then, Equity for Punks has become the world's largest equity crowdfunding raise, with over £75 million returned to the company to help it grow for our community, that (at time of going to press) numbers over 135,000. This story isn't just limited to our brewery, however. Others are growing their businesses in the same way, with equally incredible results.

In Leeds, UK, Northern Monk Brewing Company raised £1.5 million from their supporters in 13 days; their campaign ended up being 296% funded. Wild Beer Co hit £1.7 million and Siren Craft Brew reached £1.2 million from an initial target of £750,000. Bristol's Left Handed Giant hit their crowdfunding target within a single hour. This is the modern age; people who like what you do are ready to back you with their own money so that you can continue.

That's just a modern facet of the greater trend within craft beer, that people are prepared to pay for quality. Offers and discounts aren't needed to get the word out. There's no BOGO in craft. Customers of craft brewers acknowledge the work and effort and understand the logistics required to get the beer to them. Whether they invest in the brewery or visit a taproom, the provenance of craft has struck a chord with beer drinkers.

We're biased, of course, but it's because those who drink craft beer are switched on. They know the score – about the product itself, but also about what it takes to make it in the right way. They know we are determined to make a difference in a world still dominated by industrially made, bland beer. It's your call what you drink – but there's a whole other alternative out there now, driven by the people who make it and the people who consume it.

Just look at wine. One of the ways in which people like to poke fun at craft beer and its community is the colorful flavor descriptors we use. Can beer be "floral?" Should it be "resinous?" Why would you want one that tastes of sour cherries? Well, the wine industry has been using flavor descriptors for years and it worked for them. Craft beer is taking lumps out of big brewers who rely on image and branding because people are learning about what is available to them – and to do that you need words like "floral" and "resinous."

The culture of modern breweries is reflected here, too. Leaving no stone unturned when it comes to experimentation, such is the spirit of co-operation within the industry that nothing stays secret for long. But also, nothing is sacred. A beer idea will spread and enable drinkers to try different versions, wherever they are in the world. Craft beer culture has gone global. You can have whatever you like, wherever you are, from a brewery you now own a part of. That's craft beer. And that is why our culture rules.

THE CHARLATANS

Everything you have read in this chapter up to this point, Big Beer knows. The hugely powerful multi-national corporations behind many of the brands you see in shops, bars and online have had enough time and enough experience of "craft beer" to plan a strategy of their own. A way to hit back, to fight a rear-guard action as sales of their monolithic lagers continue to dwindle. To do so, hey have come to a realization: they have none of the community, openness, and experimentation that craft beer has. Industrially produced beer is churned out, around the clock, to taste as similar as possible to every other batch ever made. Large-scale breweries have only one thing on their side – bottomless coffers with which to push their homogeneous product. So, they use their resources to help their beer the only way they know. They play shifty.

They tend to do this in one of two ways:

The first is the most simple. From on-high in anonymous glass boardrooms, executives of these companies simply flash the cash and offer to buy a craft brewery they have identified as willing to sell out. This is the Route One method. If you can't beat them, own them.

Breweries decide to sell to larger concerns all the time; they have done ever since brewing became an industry. It's a part of our life. It's also an effective means of eliminating the little guy; swallow them into your operations and move on. It results in a few people in those breweries getting paid and a lot more getting screwed. Everyone at a craft brewery that is bought out by an industrial brewer has to sit down and take stock – do they stay or do they go?

THE BIG BOYS KNOW WHAT THEY ARE DOING.

Often, employees at the smaller brewery – particularly if they are founders – are made to sign agreements that they won't leave for a certain number of years. The big boys know what they are doing. Buying a craft brewery means acquiring a known quantity. Customers of the brewer will be on the lookout for them to jump ship at the first opportunity and start again. So they keep them where they are.

The second method increasingly used by Big Beer to snatch market share back from craft is to muddy the waters with pseudo–craft brands. "So-called crafty beers." Brands that are developed to look, read, and feel like craft beer, released to pick up traction and do well. But they are Fake News. These beers are charlatans that impinge on the blood, sweat, and tears of those of us trying to do it the right way.

These brand-a-like beers are typically promoted by the mega-breweries using the same wad of cash they use to buy craft producers. Heavily advertised, in an industry that doesn't usually self-promote, they appear on shelves in the same radius as craft beers and to the uninitiated don't appear to be different. It's the cuckoo in the nest, feeding itself on the back of the hard work of others.

And the truth of it? Most of these beers don't even cut it. You can tell, if you ever drink one. They are brewed by a company that is controlled and structured. Every X and O on the balance sheet accounted for. To create a beer with the freedom of a craft brewer under those constraints is actually incredibly difficult. And it shows. These beers look the part, but fall way short of the real thing.

So why is this a problem?

Well, the more craft brewers that are taken over, and the more "crafty" beers that end up in your shopping basket, the more the craft beer industry is devalued. If you pour away a beer, would you go back to the same shelf next time? Would you bother learning more about brewing? Would you tell your friends? No, you'd stick with the same lager you always drank. And the big boys win.

Not on our watch.

THESE BEERS ARE CHARLATANS THAT IMPINGE ON THE BLOOD, SWEAT, AND TEARS OF THOSE OF US TRYING TO DO IT THE RIGHT WAY.

THE PIONEERS

ARMAND DEBELDER

LOCATION: BEERSEL, BELGIUM
RENOWNED FOR: BLENDING AND BREWING LAMBIC

Adding the boiling wort into the coolship at Drie Fontenien, Beersel.

"Everything I hear is barrels and bubbles," says Armand Debelder with a shrug, as he stands outside his brewery in a small town near Brussels. As he says this, his hand is held about waist-high from the ground. He's describing his early years, growing up around the family business that was established by his father Gaston in 1953. Since he was up to his father's waist, Armand's life has revolved around the brewing, blending, and fermentation of lambic; it formed the soundtrack to his early years.

"Children didn't follow fathers into business," he says, as the sun streams down on the small yard just opposite the train station where people are waiting for buses to their Sunday-afternoon destinations. "But I started believing in myself. Saturday, Sunday – it was the work. I could go out to any time on Saturday night – but at 9am I had to work." He hits one hand into the palm of the other. From a young age, he was immersed in the way of life of a family that blended lambic.

For much of their history, Drie Fonteinen were a "geuzestekerij", a place where – rather than brewing their own lambic – the brewery blended those from others. Acting as the repository for lambic breweries without their own barrels, Drie Fonteinen safeguarded the work of others, using the skill of Gaston (and later Armand) to continue to let the beer develop in the wood and then blend the different barrels to the specifications of the brewery. This is an industry that is quintessentially Belgian. The rest of the world has contract breweries (they have them in Belgium, too, of course) – but these are "contract conditioners".

Beersel sits in the Seine Valley, the river flowing just to the west of the town; the center of the lambic region. Back in the fifties, there were 14 different blenders just in the one town of about 20,000 people, until they began to close and Drie Fonteinen were proposed as the "regional blender."

"THE BEER IS THE STAR...I AM NOT. I WILL NOT BE THE STAR."

Armand admits that he was daunted by the reputation of his father, when he took over the business. Although he had stepped down, it was clearly Gaston's realm – Armand describes him as the "*padre familias,*" looking down on everything that he did. "He only ever gave me one compliment, his entire life," he says. "'Armand,' he said to me one day, 'You never have to change what you are doing anymore.' And that was it." An acknowledgment from his father that he had finally mastered every aspect of the blending process.

Without a son of his own, Armand has passed on the torch to a young brewer named Michaël Blancquaert, and this "next generation" continues with the day-to-day operations. Armand, now in his mid-sixties, is slowly taking a back seat, having been persuaded by his wife to do less around the brewery. Particularly since the brewery has moved away from simply blending other people's lambic to creating their own (brewing started at Drie Fonteinen in 1998).

The fire still burns, though. He mentions brewing giant InBev in passing and immediately sucks air though his teeth like a hex, sounding like a hissing cat. Regarding their tactic of taking over other breweries, he says, "When you've worked all your life, you can't see that passion go and be left only as a name, if that." Armand is obviously not interested in handing over anything – you get the impression he wouldn't do it, simply for the sake of his father, if nothing else.

"IF YOU PROMISE US YOU WILL CONTINUE, WE WILL HELP YOU."

It has not been smooth sailing, however. With a series of facilities for brewing, blending, and storing their beer, in 2009 the entire business was put at risk by something as small as a faulty thermostat. Gueuze re-ferments in the bottle at between 60–64°F (16–18°C). In May of that year, an electrical fault removed the pre-programd upper limit at Drie Fonteinen's warehouse and let the hot air blower run uninterrupted. Temperatures in the building rose to 140°F (60°C) and the build-up of pressure caused by overheating resulted in 13,000 bottles exploding. In all, the spoilage accounted for a total of 80,000 bottles. With no insurance – and with beers created by other breweries lost as well as their own – the accident nearly finished Drie Fonteinen overnight. "But everyone helped," Armand recounts with a rueful grin. "'Armand,' they said, 'if you promise us you will continue, we will help you.'"

Local brewers rallied to the cause, producing one-off beers that raised money ("many of which ended up on eBay," he sighs). Others donated funds or manpower. The surviving bottles from the warehouse were sent to a distillery and concentrated down to the 40% Armand'Spirit (which proved so popular, it is still available to this day).

You can tell that he was genuinely amazed by the level of support, as people helped Drie Fonteinen get back on its feet – even if it was 2013 (when the brewery moved to its present location) before they began brewing again. Since then, Armand and his team have only gotten stronger. In 2016, they moved much of their operations, including the bottling line, to a new warehouse in Lot, less than 2 miles (3 km) west of Beersel.

Throughout all of the ups and downs, Armand has never lost sight of what is truly the most important factor, everything else being equal. "The beer is the star," he says simply, pointing at his oak casks. "Not us. People ask me to be the star. I am not. I will not be the star."

THE PIONEERS

LAUREN LIMBACH

LOCATION: FORT COLLINS, COLORADO
RENOWNED FOR: WOOD AGING AND BLENDING

"I think about them all the time," says Lauren Limbach, talking about the 65 foudres she oversees at New Belgium Brewing Company in Colorado. "Oversee" used to be in both senses of the word, but not anymore. "I moved my office. It was upstairs; I had a door, a window, a carpet. One day, I just felt so disconnected that I moved my desk right in the middle of them. And that's where I sit. If I'm cold, they are cold. If I'm hot, they are hot. Not a week goes by without me being there."

The giant oak vessels slowly age their precious contents at New Belgium Brewing Company, Colorado.

The stewardship of these silent oak vessels and their cargo of slowly aging beer, is clearly much more than a job for Lauren – it is a passion that has developed over the course of her twenty years in the industry with New Belgium. But it wasn't always the case – she found her place among the giant foudres through a different passion. Snowboarding drew her from her native Georgia to the Rocky Mountains, back in 1995. It was the slopes she wanted to be near, not barrels. "I didn't drink beer at all, before that. Never even considered it."

"EVERYTHING I DO IS RESEARCH AND DEVELOPMENT."

In those days, locally brewed beer was dwarfed by the mountain-sized shadow of domestic lagers. The city of Fort Collins had two main independent producers – Odell and New Belgium – and it was an amber ale from the latter that changed everything for her. "Fat Tire was everyone's first craft beer in Fort Collins," she continues. "You drank it because it meant something to you [being brewed locally] even when you were still wrapping your head around it."

New Belgium Brewing Company had started life a few years earlier, in the summer of 1991, with cofounders Kim Jordan and Jeff Lebesch creating Fat Tire alongside a Belgian-style Abbey Ale. After quitting their jobs, taking on loans, and scaling up their company from home-brewing in a basement, the brewery gained a loyal following in Colorado through sheer hard work. Kim would load up her station wagon with the beers and drive from bar to bar to restaurant, introducing it and letting it speak for itself.

The brewery reached capacity almost immediately and never stopped. "The growth was exponential," remembers Lauren. "Fifty percent; one hundred percent; again and again and again. It was that moment – that spark – at that time, this was all brand new. It was the craft brewers bringing this palate of flavor to everyone who had literally never had it." Imported beer from Europe was the only outlet before US-brewed craft beer appeared, and much of this arrived past its best and at a price point that was prohibitively expensive.

As a case in point, American-brewed Belgian-style beers were so rare that, when Abbey Ale gained enough of a following that New Belgium entered it into 1993's Great American Beer Festival, it fell into the catch-all "Speciality Beer" category. So few were seen that the organizers didn't have a group for them (unlike now, where the festival has over two dozen Belgian-style categories). Abbey won a Gold Medal nonetheless, becoming champion Speciality Beer.

Fast-forward a few years to the very end of 1997, and Lauren turned her discovery of Fat Tire, Abbey Ale, and other craft beers into a career at New Belgium. She joined the quality team and began pioneering the use of sensory science and taste panels to rigorously assess all of the beer leaving their Fort Collins location. This was her first love, and it's clearly gratifying to her that so many other breweries these days are putting their faith (and their money) into the sensory sciences.

"New Belgium put a huge amount of love and energy into sensory," she says. "You know, beer is only supposed to be a product that is meant to be enjoyable, and that is such a subjective and unobtainable word that for the longest time people just "liked" it. You make a beer to a standard and if it's not there you fix it. But sensory science is so much more. It covers everything from raw materials to final packaging. If a beer tastes of banana then you know you've got a yeast problem – sensory is about these giant clues that are almost begging you to find out what went wrong."

Lauren's bread and butter was the rigorous testing of Fat Tire and New Belgium's other regular releases. Increasingly, however, her quality-related role gave her a chance to assess the work going on in the brewery's small store of barrels. "I would taste them all, and the foudres as we got them. This started developing my protocol of tasting and blending. But with my full-time job running the sensory panel and training panelists I had to do it in almost a vacuum."

She would come in at weekends and at night to taste the beer that was aging. "I did it on my own, when no one was there. Even though the wood cellar grew around us, it was always the thing we did in our spare time." Lauren worked alongside her now ex-husband Eric Salazar and ex-Rodenbach barrel master Peter Bouckaert in breaking new ground in Belgian-style brewing and blending in the United States. Twenty years on, she is still there, amid the forest of foudres.

"SENSORY IS A SCIENCE, BUT... IT IS SORT OF AN ART, TOO."

"Sensory is a science," she continues. "But when you walk into the wood cellar, it is sort of an art, too. Some of these beers can take years to create. The raw materials, the barrels. I once looked for golden gooseberries for three years. Everything I do is research and development. And that's because everything that I make, I am usually trying something for the first time. But whenever I make a beer I already know what it's going to taste like, look like, and smell like. I just reverse engineer everything. I go back and make it happen."

This foresight has put Lauren at the forefront of the industry, and a few years ago New Belgium responded in kind, making her their Speciality Brand Manager. She flew all over the world, educating and explaining the art and the science of what she does. But eventually, she returned to the wood cellar and it became her sole focus again. You get the feeling this is her comfort zone – or, as she puts it, "if you're one step, two steps away from it, you lose all touch of how to do it really well."

Lauren has the rare ability to mentally keep track of everything developing inside the 65 foudres, and at any one time project this into the future. It's like working an air traffic control station but also keeping track of planes that will be flying in a year's time. "I never acted like I know what's going on in the barrels," she concludes, with typical humility. "I just do a great job of listening to what they need. I'm there for them. I'm their caregiver. Their guardian." Clearly, New Belgium is in great hands.

THE PIONEERS

SEAN FRANKLIN

LOCATION: HARROGATE, YORKSHIRE, ENGLAND
RENOWNED FOR: PIONEERING THE USE OF AMERICAN HOPS IN THE UK

When you put it to Sean Franklin that he was a pioneer, you are met with a self-effacing chuckle. "Well, it's very kind of you to think so," he says, as he recounts the story of how the two pivotal breweries he founded began with a study of French wine and ended with his retirement from brewing in 2011. Over the intervening decades, he had a number of eureka moments and changes of outlook that laid the groundwork for hop-forward pale ales to take off in the UK.

That's not hyperbole, either. As the 1980s began, his research into the flavors and aromas that could be wrested from American hops led him to start importing Cascades and Chinook from the Yakima Valley in Washington State – by his reckoning the first British brewer to do so. At that time, emergent craft brewers such as Ken Grossman at Sierra Nevada were exploring the use of Cascades for their pale ales, but nobody was doing the same in the UK.

As a result, piles of American-grown Cascades became his for the taking. "The character of these hops fitted the bill for what I wanted to do, with beers that were more floral and fruity. But demand for the hops was so low; you could get fifteen samples of one hop sent directly to you [from America]." There was practically no competition for their use in the UK, so Sean filled his boots (well, his brew kettle), using them in a brew for the first time on August 18, 1981.

The beer he created was Franklin's Bitter – one of the most formative in modern British brewing history. Deploying American-grown Cascades alongside English Fuggles, the hop dose made perfect sense; in the hop fields of North America the former were derived from the latter. They gave his eponymous debut bitter an orange citrus note that paired perfectly with the toffee flavor that emanated from the crystal malt also featured in the recipe.

His ability to marry flavors came from his background in wine. After studying oenology in Burgundy, he gained a Diplôme Universitaire d'Aptitude à la Dégustation (or DUAD) and returned back to the UK armed with this high-level tasting qualification. It led him to a Quality Control post at a company importing French wine to be packaged for the UK market. However, when the UK joined the Common Market in the mid-1970s, their insistence that all wine be bottled at source swiftly put put that career to rest.

As he admits, he was "at a loose end" and began driving a taxi to pay the bills, until a moment of clarity arrived, as they all do, without warning. "One Sunday morning, I was at a pub in Goose Eye near Keighley and I had a bright spark," he describes. "Sitting in bright sunlight, in a pub I hadn't visited before, I had a half of a beer that instantly made me think *'I could do this.'* Eighteen months later, I had my own brewery."

IF YOU HAVE AN IDEA, YOU CARRY IT THROUGH.

When you express surprise at the speed with which it happened for him, he proves how driven he was. "It gets you going, doesn't it? If you have an idea, you carry it through." In the meantime he started home-brewing, although by his own admission, "of the forty or so home-brews I made, only around five were drinkable." He learned on the job, and contacted other brewers and yeast companies. He read books on brewing from the 1940s and '50s.

HIS BIGGEST LEGACY, WHEN ALL IS TAKEN INTO ACCOUNT, IS UNDOUBTEDLY HIS FORESIGHT.

The next problem was where to put his fledgling brewery. Sean knew the area around Harrogate and Knaresborough in North Yorkshire, so headed there – the former had between 20 and 30 pubs. "The issue was that only two of them were free houses that could keep the beer if I made any," he continues. "One of them did have a barn. The Gardener's Arms. So I converted that and put in a seven-barrel brewery."

What condition was the barn in when he first set foot inside? "Oh, it was in a disgraceful state. It had a mud floor and barely a ceiling. There was a conservation order on it. But in six months I had the brewery in there, with the help of some guys from the local branch of CAMRA [the Campaign for Real Ale]. I did the drains myself." This brick barn, with two small high windows and a sloping tiled roof became the home of Franklin's Brewery – and Franklin's Bitter.

"I remember a time I was in The Junction in Otley," Sean says. "I had a pint of Timothy Taylor's Landlord. The brewer then, Alan Hay, had won competitions with it and it tasted of tangerines and oranges. It blew my mind." He started using those American-grown and -harvested Cascades and Chinook in place of UK-grown hops for his bitter, which proved an eye-opener for many people after a pint in mid-'80s Yorkshire.

"I soon ran into the problems that small brewers face, but also I was right at the beginning of the market. It was really hard to sell in a Yorkshire town with only two pubs open to me. My production was low, but even so. Eventually I made the decision to sell, and after four years, when my rent had risen to more than I could manage, I did just that." Talking to Sean, you realize that he had regrets about this almost right away, even if it was the only decision he could make.

He worked on getting another brewery, to refill that part of his life, and eventually did so through a friend of his out in Seattle, presumably made through his trips to the Yakima hop fields. The beer writer and columnist Vince Cottone, who many believe first coined the term "craft brewery" at around the same time, knew of second-hand brewing equipment and Sean managed to buy what he needed and have it shipped to Yorkshire.

This was his second coming, the arrival in 1993 of Rooster's Brewing, named after Rooster Cogburn from the Western True Grit. Sean set out from where he had left off, brewing hop-forward pale ales for a market that had started to understand what they were about, if not exactly snap his hand off to order them. He still used the American hops more than anyone else on the British side of the pond to popularize them, and persisted until they found a new audience.

He also decided to see their source for himself. "I went out to Yakima and found another world completely. The selection, the harvesting, the hop farmers who have a month to secure the future of their business for another year. The respect I have for them is so high; the stress they must feel I just couldn't handle." Starting your own brewery from the ground up, twice, isn't exactly for the faint of heart either, though.

Rooster's went from strength to strength, and expanded several times. In 2011, however, he called it quits, nearly 20 years into his second stint as a brewer, and sold the brewery to the Fozard family. They still run it today, having expanded again and continue to produce world-class pale ales brewed the way Sean would have wanted. To celebrate his quarter-century in the industry, the new Rooster's produced a Cascade and Chinook pale ale in 2018, naming it "Franklin".

This is just a small indication of the debt modern craft brewers in the UK owe to Sean. From his pioneering importing and use of unused hops, to his reliance on a background in wine and its associated flavor expertise, his dedication even led to him delivering casks from the trunk of his taxi (when asked if he had passengers on the same ride, he politely says he can't remember; smaller firkins maybe, lager kilderkins would have meant no room on the back seat for people).

His biggest legacy, when all is said and done, is undoubtedly his foresight. As he summarizes: "My past in wine gave me an idea of what I thought would come to the beer trade, and it has. The floral and fruity beers arrived and now there are so many more. Although now they are more like fruit juice." So, does he still have the bug? "Oh yes. The bug ticks away. Good beer is so much more widespread now. I'm not going to break out and go back into it, though."

Like any pioneer, he has earned his quiet time of retirement.

Innovative blends of hops are Sean's trademark.

THE PIONEERS

JASON PERRAULT

LOCATION: YAKIMA VALLEY, WASHINGTON
RENOWNED FOR: GROWING AND BREEDING HOPS

Jason Perrault's career in hops started early. Growing up on the family farm, he began helping out by rolling twine at the age of five. "My father would set us kids up in the backyard with a bundle of twine and we would sit on lawn chairs rolling it up," he explains. "The job was to wind it around your hand into the shape of a doughnut. I think he paid us something like a penny a roll. It was a pretty easy and simple job, but a good way to begin teaching kids the value of work at a young age."

Anyone who grows up in the shadow of a family business can tell you similar stories of pitching in at a young age – but the Perrault hop farm has youthful determination at the center of its story. In 1902, Jason's great-grandfather Alberic boarded a Northern Pacific Railroad train from Minnesota with a single, but daunting, duty – to establish a homestead for himself and his family, somewhere in the wild west. He was 18 years old.

Following the advice that had echoed in the ears of a previous generation, Alberic did indeed "Go West, Young Man," and together with his wife Mamie created a homestead in a valley near the town of Moxee, Washington. Recent irrigation had been installed there to attract new farming families, many of whom were French speakers (mass at the church in Moxee was given in French until the First World War).

THE PERRAULT HOP FARM HAS YOUTHFUL DETERMINATION AT THE CENTER OF ITS STORY.

Alberic, Mamie, and their 13 children became hop farmers in 1928. Today, their farm remains renowned for the breeding and growing of the same crop, four generations on. A quick glance at some of the varieties developed by companies with links to Perrault indicates just how much: Citra. Mosaic. Simcoe. Ekuanot. Ahtanum. Warrior. Palisade. Loral.

Perrault Farms was a founding partner of Yakima Chief Ranches (previously known as the Select Botanicals Group), along with a pair of other hop-growing family farms: the Carpenters and the Smiths. In 2003, the Yakima Chief Ranches breeding program joined with local hop supplier and breeder Haas to form the Hop Breeding Company (HBC). All resources and breeding efforts were pooled, and all brands that resulted were released jointly.

The result was a boon for commercial brewers, as the Pacific Northwest became synonymous with pest- and disease-resistant hop varieties that arrived courtesy of a single supply chain. And the most important link in that chain – although he may modestly say otherwise – would be Jason, and his encyclopedic knowledge of cultivating hops. Although, in all probability, Jason would likely say that the key link would be where all the growing takes place – the Yakima Valley itself.

Perrault hop farm, Yakima, Washington.

"We have the best environment in the world [in which] to grow hops," he says. "The Yakima Valley is a high desert with fertile volcanic soils and 300 days of annual sunshine. Our latitude ensures the proper daylight, yet our winters provide adequate chiling hours, making for a well-balanced growing season. We have a sustainable source of water in the form of run-off from the Cascade Mountains, stored in reservoirs and snowpack and then delivered via an intricate system of canals."

Irrigation is what first attracted the hop farmers to the region and it remains crucial today. That's because the Yakima Valley receives only 7–10 in (18–25cm) of rain a year (compared to 30in/ 75cm in London, 48in/120cm in Sydney and 90in/ 230cm in Singapore). Even Seattle, just 130 miles (210km) away from the modern-day home of Perrault Farms in Toppenish, receives 40in (100cm) of rainfall a year. Hops need water – and the hidden canals of the valley bring it to them.

The dry, arid conditions reduce the impact of damp-loving diseases such as powdery and downy mildew – and the cycle of the seasons is perfect for the hops; the long days of sunshine arrive at the right time for growth, and the short nights follow at the perfect time to induce flowering. That said, Jason is a farmer and bad weather is always on his mind – hail is his biggest concern. Likewise, heavy rain followed by high winds, which can bring down entire rows, with telegraph-pole-sized anchors bowing inward until they snap.

When you visit Jason and walk the hop farm with him (always accompanied by his field dogs Hammer and Brady), you realize the scale of what he has to process on a daily basis. Particularly in the experimental fields – row after row of single plants or small groups solely identified by a three-digit number; the next Mosaic or Centennial among them. He keeps track of many thousands of different varieties, crossing them together in a never-ending search for citrus, berry fruit, or pine (or anything else you can think of).

It wasn't always this way. A decade (or more) ago, brewers would arrive at Perrault and ask for hops that were all alpha acid; they only sought bitterness. Low-aroma was the order of the day. "This was based on the styles of beer being brewed," Jason explains. "The prominent buyers of US hops at the time were large multi-national brewers making lighter lager beers. Their focus was on efficiency in bittering versus high-impact flavor. The rise of craft beer and IPAs has changed all that."

And how. These days, there are developmental hops with descriptive traits like bubblegum and piña colada; colossal grapefruit bombs and multi-layered stonefruit alternatives; many hops with highly specific flavor attributes. But therein lies the rub. As Jason puts it, in order to be a success, these experimental varieties need to walk the walk and "translate consistently and stably into beer and enhance the complexity."

This is the Holy Grail – getting those "wow" moments the brewers experience in the hop fields into a beer and passing them on to the drinker. It is why experimental hops take time and sheer hard work. "The breeding process takes about ten years," continues Jason. "This process is simply a series of vetting steps to take a starting population from thousands of individuals down to just one or a few with the right combination of characteristics."

If they feel a strong pull from the brewing industry that a new experimental hop is working, Jason and partners at YCR and HBC consider releasing it as a new commercial brand. But even that can take a while to land. Take Simcoe. Now a household name (and a big part of the dry-hop regime of Punk IPA), as Jason says "when YCR first commercialized it in 2000, we could hardly give a pound away. It was simply not the right place in time for it."

Simcoe was too punchy for the millennium-era beer scene. The three partner farms of Yakima Chief Ranches were the only people to grow Simcoe for the first eleven years of it being commercially available. However, in 2011, its use suddenly spiked with craft beer's continuing trajectory. Between then and 2019, those initial three farms have been joined by around 70 others, with over 40 Simcoe-growing hop farms in the Pacific Northwest alone. It is now the third most widely planted hop in the United States.

Simcoe paved the way for all of the unique aroma hops that followed, and so too – of course – did the Perrault Farms themselves. From the days of the pioneers out west to their continued pioneering work developing the hops of the future, beer drinkers everywhere owe a huge debt of gratitude to Jason and his family firm. And it all began with an 18-year-old boarding a train to an unknown destination.

"WHEN YCR FIRST COMMERCIALIZED [SIMCOE], WE COULD HARDLY GIVE A POUND AWAY."

THE PIONEERS

EVIN O'RIORDAIN

LOCATION: LONDON, ENGLAND
RENOWNED FOR: BEING AT THE FOREFRONT OF LONDON'S BEER SCENE

If craft beer is currently blazing a trail, then one of those fiery arcs surely emanates from the UK's capital city. It's fitting, as London has always been a brewing town; from the first inhabitants, through the trading days as its ports opened up the world, and on to the working porters who fostered changes to entire families of beer in their downtime. But the recent stories that pour forth from either side of the Thames are increasingly unique, innovative and… well, simply increasing.

As the London brewery count rises well into three figures, it must be increasingly difficult to stand out – to separate yourself from the noise. Operating in a teeming market has its positives and negatives, so maybe the best way to be is as unperturbed as possible. Or, to sum it up in less than 20 words – "We quite like the beer we make. We don't feel the need to follow styles or do other things." You stay relevant by being true to why you started.

If you have even a passing experience of London beer, it will come as no surprise that this most philosophical of philosophies comes from Evin O'Riordain, founder of the Kernel Brewery. Since they opened the doors of their initial 140-gallon (650-liter) brewery in September 2009 under a railway arch in Bermondsey, so much has changed – and yet Evin and his small team (a decade on, their company has grown to 14 people) continue their path, as laid-back as they can be in a global metropolis.

Maybe laid-back is the wrong choice of words – confident would work, but nothing about the Kernel is as over the top as that word sometimes suggests. It's hard to pigeon-hole them. But with the Kernel you really don't need to. Few other breweries let their beer do the talking as much as they do, bottles still sporting the iconic brown paper labels.

"SETTING UP YOUR OWN THING… WELL, THERE'S NOTHING BETTER."

"Everything we built in was done to make the beer better," continues Evin, talking about how the brewery grew in the early days. "We could have gone with more automation, could have added a hop doser, for example. But we wanted to retain that manual connection. We still dig out the mash tun and the copper by hand. We still chuck in hops ourselves. It's how we relate to what we do. We currently have 14 staff; 12 in the brewhouse. We all take turns brewing."

That number, as small as it is, does change, though. Not long before our conversation, one of Evin's longest-standing brewers left to move to Scotland and start his own brewery, which obviously brought out mixed emotions. "I mean, it's brilliant – we miss them, but doing your own thing is far more preferable. The world would be a better place with more small breweries. In terms of your health as a human being, setting up your own thing… well, there's nothing better."

The Kernel started life as a 650-liter brewery back in 2009.

For Evin, it was a visit to New York in April 2007 that led him to do just that. At the time, he worked as a cheesemonger for Neal's Yard Dairy; a forerunner of the British wholefood movement, founded in 1979. During his few days on the Lower East Side, he was tasked with showcasing cheese to the staff of a Whole Foods Market. At night, as he puts it, they "turned the tables" and taught him about beer, something he had never considered.

"I drank beer, of course, but until that point I'd never given it a thought. The US were ahead of the curve, and the beers there made me go 'Aah, beer *is* worth thinking about'. This was over ten years ago; there was no culture of talking about beer – it was there to lubricate social interactions in the pub, as it always will, which is fine, but the US enthusiasm and range of simple hop-forward pale ales ... that was it." Evin returned to London and started home-brewing.

He did this for a "couple" of years before opening the Kernel brewery, just to the south of The River Thames in Bermondsey. Now a place fundamentally associated with craft breweries, back then it was very different; as were the beers he compared what was around at the time – in his own words, directly flavored, clean, simple hop-forward pale ales with low yeast character. All these years later, these keywords are still all directly applicable to the Kernel's beers.

There were digressions along the way. Evin chuckles as he recalls the "London Murky" years around 2010 when the Kernel released IPAs with heavy sediment in the bottle, frequently ending up transported and poured as cloudy as modern-day New England IPAs. "Those are the only times I feel old," he ruminates. "When we did heavily hopped, cloudy, juicy IPAs [back then] and everything has gone past to where we are today. Our beers have less sediment today, but they have stayed relatively consistent."

This is about as big an understatement as you can expect to hear – the Kernel is known for consistency. They still bottle-condition and keg-condition all of their beers, even with reduced murky results. Evin ponders the right phrasing to use, when considering why they still condition all of their releases in this way, before saying, "Let's call it a soul. Bottle-conditioning is part of our DNA; it gives texture and life to beers. Secondary fermentation makes them what they are. If we carbonated, or put our beer in cans, I don't think they would survive the transition."

Although the Kernel is consistent, they are not afraid to change things up. In 2015, they closed their hugely popular taproom after it inadvertently kick-started the phenomenon of the "Bermondsey Beer Mile" – a run of breweries open at weekends that became a crawl. "It was intense," Evin says, reliving the peak days. "Everyone was crammed into the warehouse, we couldn't deal with the space. We had people waiting 20 minutes to get a pour. It eventually became somewhere even we wouldn't want to visit."

The decision was made to close the taproom – but four years later it has reopened, entirely on their terms. Their mission is to create atmosphere and move away from the constant beer festival vibe. The key, they have realized, is a separate space. As Evin sums it up, "people want to have a beer. Why make it difficult for them?" He says later that there's never been a better time to be a beer drinker, and he's right – and the conversation turns to the sheer amount of choice that now exists.

The Kernel, famously, lack a core range of beers. They have a regular set of styles – pale ales and IPAs, historic London porters and stouts, and increasingly mixed fermentation beers – but they are constantly changing. "Not having a core still works," he says, when pressed to quantify where the Kernel fits with other, more recent, breweries who have the same outlook. "It works quite unintentionally, though. Many people now want a new beer every week, which is fine."

"But our pale ale is our pale ale," he concludes, almost ruefully. "If the beer world is always chasing the new, that's [also] fine, but it must be a hard thing to maintain the energy. Our brown labels are removed from that world. They stand out against the color, even as the rest of the world has gone technicolor." The Kernel are conclusive proof that you don't need to be loud to stand out.

"WE QUITE LIKE THE BEER WE MAKE. WE DON'T FEEL THE NEED TO FOLLOW STYLES OR DO OTHER THINGS."

THE PIONEERS

SIR GEOFF PALMER

LOCATION: EDINBURGH, SCOTLAND
RENOWNED FOR: PIONEERING BREWING SCIENCE AT HERIOT-WATT UNIVERSITY

In March 1955, Godfrey Henry Oliver Palmer stepped off a boat onto Liverpool docks, having crossed the Atlantic to England from his native Jamaica. He made the crossing alone, at the age of 14 years and 11 months. His mother was, at the time, living on the Caledonian Road in London and worked as a seamstress, spending £86 to bring her son over to join her.

As he recounts, soon after arriving in London, "a man appeared at the door" and informed Mrs. Palmer that as her son was under 15, he would have to go to school. "My main education was going to church," he says, indicating just how alien his new surroundings must have seemed. "I had no formal education and was rejected by a few schools before I managed to get a place at Highbury Grammar thanks to my cricket."

When you ask Sir Geoff if everything that then happened was the result of him earning a cricket scholarship, he laughs – because, eventually, that just-under-15-year-old garnered a doctorate and professorship, along with six honorary doctorates, an OBE, and, in 2014, a knighthood for his services to human rights, science, and charity. Oh, and he also had a pub named after him.

"It was called Palmer's," he recounts, "On Lothian Road in Edinburgh. It was owned by Scottish & Newcastle and had been called Shakespeare's, but I was then teaching at Heriot-Watt on Chambers Street and they named it after me. This would have been in the 1980s."

Did he go there? "Oh yes!" he replies, "I did the opening. I used to sit there and feel so proud of myself. But it got so popular that I think they struggled, so they changed the name back [to Shakespeare's]. He's more important than me anyway. I should think more people have read *Hamlet* than *Cereal Science and Technology*."

This entire anecdote, like many Sir Geoff tells, is punctuated by repeated laughter coming down the phone line. He then admits he feels pride walking past the pub, even today (it still exists, still named after the other guy).

"TECHNOLOGY IS SCIENCE THAT WORKS."

Sir Geoff is one of the pioneers of brewing science, having founded the world-renowned International Centre for Brewing and Distilling at Heriot-Watt University, in 1989. A generation (and now a second generation) of brewers have learned their trade there, including our cofounder Martin Dickie, who Sir Geoff remembers teaching, at around the turn of this century.

After retiring in 2005, he gained Emeritus Professor status and returns to give an annual lecture to the students who still read his research to this day, to assure them that he is still going. "They think I'm dead," he jokes. The institution he helped bring to life has, in turn, changed the lives of hundreds of brewers (and distillers) since the early 1990s – it's no exaggeration to say that without it, and him, there would be far fewer breweries out there.

Amid of all his achievements, you sense that this is the one that matters most. "I get the greatest thrill," he explains, "when I go into supermarkets and buy 'my own' beer. I'm so proud when other people buy beer from people I taught. I buy them and I assess… I can taste the malt and hops that I taught them about." This must be amazing, to be given a continual link back to ex-students in this way.

Although he mentions both of the key ingredients in beer, it is malt that Sir Geoff is known for around the world (he later says, jokingly, that "for my sins, I taught them about hops as well"). His early schooling led to an interest in botany, which then developed into a junior lab technician's role at Queen Elizabeth College, London, under Professor Garth Chapman.

"I'M SO PROUD WHEN OTHER PEOPLE BUY BEER FROM PEOPLE I TAUGHT."

It was Professor Chapman who gave Godfrey Henry Oliver Palmer the sobriquet by which he is known to this day. "I told him my full name and he looked at me and said, 'Can I just call you Geoff? If I can call you Geoff, you can have the job.'" So, Geoff Palmer started out in science and never looked back. In the early 1960s, he moved to Scotland and turned his knowledge of botany into a lifetime's work on barley.

He studied at Heriot-Watt University under the world's first female Professor of Brewing and Biochemistry, Anna Macleod. Professor Macleod (who died in 2004 and has a hall of residence named after her at the university) taught him about brewing and distilling. He worked on Proctor barley, a spring malting variety that is one of the parents of the mighty British malt Maris Otter. Or, as Sir Geoff puts it, with typical humility: "I came up with a few things."

Did he ever. Sir Geoff pioneered the field of barley abrasion, working on how barley turns into malt. At the time, it was thought the inner part of the barley kernel, the germ, was solely responsible for enzyme production – but he posited that the bran (the fiber-rich outer layer) also was a major enzyme source, and if this was true, it was possible that a grain could actually malt from both ends, speeding up the entire process. He took a pin and scratched away at the back of a single grain, before applying the hormone gibberellic acid at the front and back. Proving his theory was correct, the grain duly germinated at both ends. His abrasion process – scarification – worked. It was replicated, and then scaled up to 10 tons an hour in an industrial machine called a Simon Impact Abrader.

Barley abraision enabled an acceleration of the entire process of turning barley into malt.

"We went down to Manchester to see one of these in action," he remembers, calmly summing up the moment his theory had become mainstream. "Then Bass Charrington had eight of them. Allied Brewers had eight or more. The maltsters were concerned it made the grain malt faster and increased production. Big brewers were abrading their own – I wasn't always that popular with maltsters, even though they were my friends." He laughs again.

When he later went on holiday to Greece, he saw a brewery using a similar machine to put into practice the idea he came up with, years earlier. What did that make him feel? "It gave me a sense of confidence in the importance of science," he replies. "Technology is science that works. It gives me pride in the power of science. It's not influenced by politics, race, or religion. It's about nature and getting things right in order to make our technology manage it."

"It's why you need to educate every child," he concludes, "as one of them may be the one who ends up curing cancer."

OUR FIVE MOST IMPORTANT BEERS

One of the biggest surprises that fell out of our decision to release the recipe to every beer we had ever made was simply *how many* of them there were. The most recent update of DIY Dog tops out at 415 unique beers. Some we make every day of the week, others are one-offs unlikely to be repeated (hello, Cranachan Cream Ale).

They are all special, of course, but as we near the 500 mark, we can look back and pick out a few that were more special than others. Our formative brews. The ones that have defined BrewDog as we made the leap from garage to lock-up to the place we are today. And it all begins, as it did back in the day, with a certain blue-labeled India Pale Ale…

1. PUNK IPA
(FIRST BREWED: APRIL 2007)
POSTMODERN CLASSIC

Beer #1 on our list was beer #1 in our history, and it's quite something to consider that the first ever BrewDog beer is still our best-seller, over a decade later. But we think we know the reason; Punk IPA is a perfect metaphor for BrewDog as a company. Back in 2007, it reflected the influences of James and Martin and the frustrations that led them to start the brewery. It was a home-brew success story that ended up launching an entire business.

PUNK IPA IS MORE THAN OUR FLAGSHIP; IT IS BREWDOG IN MICROCOSM

Crucially, Punk IPA also evolved as BrewDog did, changing recipe in 2011 to gain the benefits of dry-hopping. This was a huge gamble – other companies have altered the taste of their main product and bit the dust as a result – but we changed Punk because we embraced technology and were able to add new skills to our arsenal. The all-new, dry-hopped Punk has since led from the front and is enjoyed around the world.

If you have a home-brew rig, you can re-create both the original and post-2011 versions with the recipes on pages 174 and 175.

2. TACTICAL NUCLEAR PENGUIN
(FIRST BREWED: SEPTEMBER 2009)
UBER IMPERIAL STOUT FOR THE DEDICATED

Just two years after we started, we reached the peak that was TNP. We have always been about pushing the boundaries and seeing where experimentation can lead – even over a decade ago, it was part of our core DNA. In the autumn of 2009, this outlook led to 500 bottles of a 32% ABV imperial stout wrapped up inside a paper bag hand-decorated with a penguin.

The strongest beer we had (then) ever made became the (then) strongest beer in the world, but it was also by far the most involved we had ever brewed. We took an imperial stout and aged it for a year and a half before we even started the process that led to Tactical Nuclear Penguin. A process of repeated freeze-distilling to -4°F (-20°C) created the mightiest of Eisbocks (see page 107) and TNP was born.

It's easy to forget now, when strong beer weighs down the shelves of any self-respecting bottle shop, but the response to Tactical Nuclear Penguin was pandemonium. It was greeted by a storm of media articles about our irresponsibility, sparking debates on binge-drinking and minimum pricing. A lot of people took it really seriously – but it was never intended that way (hence the paper bags). Many of the conversations it started are still going today.

3. DEVINE REBEL (W/ MIKKELLER)
(FIRST BREWED: DECEMBER 2008)
COLLABORATIVE OAK-AGED BARLEY WINE

If the theme of these pivotal beers is things that run through our DNA, then alongside the boundary-pushing that led to Tactical Nuclear Penguin we have to include the spirit of collaboration. Over the years, we have proudly worked and brewed with dozens of leading breweries from around the world – some big, many small, all independent. These are brewdays that get highlighted on our calendar.

They all started here, in the freezing Aberdeenshire winter of 2008, when Mikkel Borg Bjergsø journeyed to Scotland for what would be the first BrewDog collaboration. A high-gravity, single-hop (Nelson Sauvin) barley wine, Devine Rebel ended up in oak and then inspired a version that we aged in a Mortlach Speyside whiskey barrel for almost two years.

Both of these beers had many secrets, but chief among them was the fermentation profile thanks to champagne yeast, which led the original Devine Rebel to 12.5% ABV and a massive, effervescent presence that lifted what can be a chewy, thick style. Mikkeller are masters at the arts of blending styles, and it's amazing how both our breweries have changed since that cold day, back in Fraserburgh.

4. #MASHTAG
(FIRST BREWED: MAY 2013)
DEMOCRATIC BREWING IN ACTION

Back in the spring of 2013, we decided to take a look at the way in which our supporters could get even closer to the beer-making process. Anyone can visit and look around our brewery to see the mechanics of it all – but there had to be a way to open up our thought-making process when it comes to new beers. The answer was a eureka moment – we let the public make all the decisions. #MashTag was born.

#MASHTAG BRINGS PEOPLE TOGETHER AND THOSE WHO DRINK OUR BEER CLOSER TO THE ENTIRE PROCESS. IT'S A WIN-WIN.

Throughout the course of a week, every year (with the exception of 2017), we throw a series of questions out to our community on the way they want the beer to go, and they answer. From the overall style, to the ingredients used, to a potential twist and then the final label artwork (which we ask them to design and then vote on). Every single aspect of the #MashTag beer is their call.

The results have been amazing – and not just when it comes to the beers. #MashTag week is a time of year we all look forward to, because we never truly know what it will bring. Every other beer we make is in our heads from day one. This beer is different – and therefore special.

#MASHTAG THROUGH THE AGES

2013: AMERICAN BROWN ALE (ROASTED HAZELNUTS, ALMONDS, AND OAK CHIPS); ABV 7.5%. LABEL ART: ROB MACKAY.

2014: IMPERIAL RED ALE (BLOOD ORANGE AND CITRUS PEEL); ABV 9.0%. LABEL ART: CHRIS BRITTEN.

2015: BLACK BARLEY WINE (OAK CHIPS, VANILLA PODS); ABV 10.0%. LABEL ART: MARK GREEN.

2016: TRIPLE INDIA PALE ALE (SOUR CHERRIES); ABV 10.5%. LABEL ART: MARTY BRESLIN.

2017: #MASHTAG TOOK A WELL-EARNED BREAK.

2018: AMERICAN PALE ALE (HIBISCUS AND YUZU); ABV 5.0%. LABEL ART: PATRYCJA SIUZDAK.

2019: RED DOUBLE IPA (GUAVA AND SWEET ORANGE PEEL); ABV 9.0%. LABEL ART: NATHAN WILLIAMS.

5. PUNK AF
(FIRST BREWED: MAY 2019)
FULL-FLAVOR, NO ALCOHOL, ALL PUNK

We have known for years that the only way in which craft beer can break into the mainstream is if everyone can find something they want to drink. When you are an alternative to the single-note orchestra of mass-produced lager, having as many strings to your bow as possible is what sets you apart. With that in mind, we are on a mission to prove that alcohol-free need not equate to taste-free.

We have had alcohol-free beer at the core of our line-up since the summer of 2009, when Nanny State burst into the world. It remains one of our best-sellers, but a decade later, we partnered it with another – Punk AF – and it has changed the game for what some within our industry term the *no and low* sector. Call it what you want, we simply wanted to change the story there, too.

Punk AF arrived as the next step in our desire to be as inclusive as possible; to brew for every taste and every occasion. We loaded it with eight different hops to create an IPA that feels like an IPA, yet one at 0.5% ABV. It brings all the attitude of our flagship, but none of the alcohol. Craft beer is a point of difference – and the underpinning of our alcohol-free beers with Punk AF is a testament to that.

WE ARE ON A MISSION TO PROVE THAT ALCOHOL-FREE NEED NOT EQUATE TO TASTE-FREE

THE STYLE COUNCIL

CHANGING TIMES

Roughly midway between Los Angeles and San Diego, lying inland from both, is the city of Temecula. Home to over 110,000 people, its official motto is "Old Traditions, New Opportunities", which also works as the ideal metaphor for what took place there in June 1994. A beer appeared that didn't just break with tradition, it shoved it bodily from the stage. India Pale Ale took an opportunity to double its appeal.

Vinnie Cilurzo, son of local vineyard owners, was at the time brewer and co-owner of the Blind Pig in Temecula. It was at this (long-since closed) brewery and tasting room that he created its first beer – Inaugural Ale. Multiplying an IPA home-brew recipe by a factor of two, the beer weighed in at nearly 7% ABV and around 100 IBUs. At a stroke, it heralded the birth of a modern-era American beer style and launched the worship of American hops and an armsrace of bitterness to boot. Double IPA had arrived.

At least, that's the generally acknowledged story. Others claim Double or Imperial IPAs appeared earlier, and Cilurzo admitted he created Inaugural Ale not out of altruism, but necessity – his second-hand equipment was untested and a mountain of bittering hops added to his IPA would cover up any faults the kit may have imparted on the beer. Brewers are nothing if not resourceful, and always have been.

No matter the case, Blind Pig Inaugural Ale was a hit. When Vinnie and his wife Natalie left the brewery to found their own concern – Russian River Brewing Company in Santa Rosa, 500 miles (800km) to the north – he adapted the recipe again. His trailblazing Double IPA eventually morphed into Pliny the Elder, a beer Natalie named after the Roman naturalist who wrote about (a plant that may have been) wild hops in his first-century AD encyclopedia, *Naturalis Historia*.

Fast-forward a quarter of a century from its genesis at the Blind Pig in Temecula, and Pliny is one of the most sought-after beers in the world. DIPAs are mainstream, and also routinely north of the Inaugural 7% ABV. The boundaries are gone, if they ever existed in the first place. The beauty of craft beer is in the noun and the verb – the skill displayed by brewers to create different beers from all manner of different processes, ingredients, and contexts.

As with Blind Pig Inaugural Ale, an M.O. typical of the scene is that many of these beers arrived without warning, following sudden flurries of experimentation. Beer styles are respected but not revered; they are – as they should be – templates. The progression of beer styles through experimentation marks one of the key differences between craft brewing and mainstream, industrial beer. Nothing is static. Not even the concept of style itself.

In brewing, "style" relates to a large range of factors used to classify a beer – including (but not limited to) its color, history and backstory, aroma and flavor. The late, great beer writer Michael Jackson brought many of the styles we brewers bend and weave to the attention of drinkers in *The World Guide to Beer*, published exactly 1,900 years after Pliny started work on his seminal tome.

Since 1977, the styles that Michael wrote about have been adapted, bastardized, and distorted because of a single desire adopted by the brewers of the world: to try something new; to see whether a recipe they worked out and worked up will come off.

For an example of where this can lead, we need only return to craft beer's poster style. As we have seen, India Pale Ale magnified into Double IPA in the mid-'90s and never looked back. Since then, every few years or so (or more recently, every few months) another variation appears on the feeds of social media. The evolution of brewing IPA can be detailed like a family tree, passing through the generations:

IPA → Double IPA → Black IPA → Session IPA → New England IPA → Brut IPA → TBC IPA

This list isn't meant to be an exact timeline, only illustrative of the way in which brewers have changed what in the 19th century was sold as "Pale Ale for the Indian Market" into something entirely different, several times over. So, these days, we have a beer that can be both pale and black at the same time, one that revolves around the notion of how many you can have in a sitting, and another that pits beer in direct comparison with effervescent, airy sparkling wines.

BREWERS ARE NOTHING IF NOT RESOURCEFUL, AND ALWAYS HAVE BEEN.

It's not just India Pale Ale that has been transformed in this way. Over the years (many of which have arrived more recently than we care to admit), countless beer styles have evolved. Whether a slight update or a milestone moment, these shifts have changed brewing for all of us. And beer drinking for all of you. So, in the following chapter, we showcase a dozen instances where beer has come of age, beginning with a blast of Nitro power…

STOUT » NITRO STOUT

STOUT

If Pale Beer has an approachability to it, Dark Beer has more of a beguiling nature. Its color hides many mysteries. Its history is just as enigmatic – stout has been around since the 18th century and wasn't always the deeply inky color we know it by today. Pale Stout was a thing for a long time (if anyone tries to tell you that Black India Pale Ale is an oxymoron indicative of modern brewing). Stout was "stout" because it was substantial. Solid. Tough.

The dark beers that became the family of stouts we know today emerged as a stouter – i.e. stronger – version of the dark beer of the day: the porter. And so, in 19th-century England, one of the trends of brewing arose – the muddying of stylistic convention (such as it was at the time). Stout and Strong Porter were used interchangeably to describe the same kind of beer.

Whatever people called it, it caught on. As porter became less popular, stout shouldered the mantle of dark beer for the masses. And it did so because of its continuing versatility. Stout has changed with every generation that has drunk it. From the name people use to ask for it, to the addition over time of ingredients like lactose sugar, oatmeal, black patent malt, and even oysters.

Just as we used IPA as a modern example of the way in which brewers change a beer based on what is available to them and in order to make it more appealing to the public, stout has already been there and got the T-shirt. Milk stouts, oatmeal stout, the dry stouts of Ireland; the lineage of dark beer has many offshoots, all of which are worth exploring. Stout is, as it always was, the people's champion.

LEFT HAND MILK STOUT NITRO (USA)

As you'd expect, craft brewers aren't about to sit around and have stout slowly grind into stasis. The canvas has plenty of room for addition. Whether adding more atypical ingredients to mirror the days of shellfish in your beer, or brewing stronger versions destined for barrels, stout is still a beer on the move. And one of the ways in which it can move is upward and outward.

Dispensing stout with nitrogen dissolved in the beer alongside carbon dioxide is nothing new, as anyone who has watched a beer commercial over the past 50 years can tell you. But the slowly settling rings of mesmeric foam are now available in more places than on the bar counter of your local Irish pub. As canning has given bottling a run for its money in modern craft brewing, so too has the appearance of the Nitro can.

Many whizz-bang canning lines use an inert gas to purge the space between beer and lid before sealing. Oxygen is, as we have seen, the enemy of beer flavor, so its removal is important. Technology now exists to blast liquid nitrogen into this space moments before sealing. As the nitrogen instantly vaporizes, it pushes the oxygen out and imparts itself into the beer, leading to a richer, creamier pour without the need for an '80s-style floating widget.

The pioneers of this were Colorado's Left Hand Brewing, who introduced the first widget-free craft Nitro stout back in 2011, and it remains "America's Stout" today. The Milk Stout Nitro pours exactly as you would expect, in a cascading torrent of layers of ivory foam, creating a bevy of roasted mocha and milk chocolate flavors, all of which are backed with a smoothness that is as rewarding as beer can get.

SEE ALSO:

LERVIG NITRO LATTE (NORWAY)

The first time Norway's finest embraced Nitro was in this coffee milk stout. A vigorous pour is followed by a classic slow settle into a smooth, sweet mocha masterpiece.

DESCHUTES OBSIDIAN NITRO (BEND, OREGON)

As befits a Nitro pour, this is a beer to linger over. With distinct tones of espresso, chocolate, and roasted malt, there is just enough hop bite to cut the velvety sweetness.

THE WHITE HAG THE BLACK SOW NITRO COFFEE MILK STOUT (IRELAND)

A bittersweet barnstormer, combining the blackness of dark malts with roasted coffee and milk sugars to push the aftertaste to new heights.

LAGER » IPL

LAGER

Lager is much more than simple fridge filler – it is a yeast-powered family of beers ranging from light to dark in color, light to deep in flavor, and light to weighty in expectation. There's a reason why 90 percent of the beer drunk on planet Earth is lager. It is reassuring, familiar, and constant in the ever-changing world we are detailing in these pages. Or, at least, much of what is widely available is.

Deriving from the German verb *largern* (meaning "to store"), these beers are fermented with yeast that works at lower temperatures than ale yeast and are then conditioned for a longer period of time. But therein lies the problem – all lagers are lagered, but not all are lagered equally. Commercial pressure has resulted in an inverse arms race as the multinational brewers churn out golden yellow examples in timescales measured in days.

This is a crying shame. The many and different beers that constitute the lager family are amazing. Bready, refreshing Helles. Toasty, breadcrusty Vienna Lagers and Märzenbier. The dried-fruit richness you get from Doppelbocks. The roasty chocolate notes that form the backbone of Schwarzbier. All this and more is shunted aside in the eyes of the world in favor of bland beers that may or may not reference (the equally great) Pilsner.

While we have argued that India Pale Ale is the poster child of craft beer, lager is the behemoth that has come to represent beer in the eyes of the world. It has swept all before it. Easy to drink and always on tap, lager isn't going anywhere any time soon. But it is starting to feel something creeping up on it. Mainstream lager sales are in decline. There are many reasons for this, but the big boys are not sleeping as well as they used to.

TEMPEST BREWING LORAL IPL (SCOTLAND)

Craft brewers are always busy. Whether you're building from the ground up or working for someone else who has, time is money. But that is even more the case when you consider what goes on in the ivory towers of multinational brewers. Here, hours are counted in millions of dollars, euros, yuan, or yen. Everything has to be quantified and acted upon. Craft brewers work every hour, but those hours are theirs. Time can be used to their advantage.

The ability that craft brewers have to set their own agenda has resulted in some amazing craft-brewed lagers. Or craft-stored lagers, to be more accurate. Tank space is still the ceiling that affects what you can brew, but if you want to create a lager, craft brewers have the leeway to do it justice; given the proper lagering period, the yeasts can work to their fullest. Industrial brewers work on the assumption that turnover rules; less is more. But with lager, more is more.

As with many things involving craft beer, more is more is the norm. Breweries are harnessing conditioning time and marrying it with another technique that needs a while to yield best results – dry-hopping. When you combine one with the other, it creates a crossover: a cold-conditioned lager lifted by the multifaceted flavor and aroma of hops. Clean, dry, and bitter, we now live in the time of the India Pale Lager.

This recent arrival is transient – many IPLs are summer seasonals, brewed, stored, and poured in perfect context. It's an irony given how they are created, but these are not beers that hang around. In the Scottish Borders, Tempest Brewing Co. created a summer IPL showcasing the Loral hop, itself a crossover between noble European varietals and American muscle. Floral, citrus zest and freshly cut grass abound in this nigh-perfect example.

SEE ALSO:

STONE TROPIC OF THUNDER (ESCONDITO, CALIFORNIA)

This looks like a regulation lager until you get your nose in. Pineapple, melon, white wine, and more. Stone call this a Tropical Lager rather than an IPL, and who are we to argue?

BRUSSELS BEER PROJECT WUNDER LAGER (BELGIUM)

At 3.8% ABV this is a picture-postcard thirst-quencher. Bitter lemon and grapefruit come to the fore and lead to a moderately bitter and hugely rewarding finish.

DONZOKO ULTRABRIGHT (ENGLAND)

Hartlepool's finest have split the IPL into its constituent parts: tropical hop-derived flavor and crisp, clean lager bitterness. This is a triumph of the juicy and the dry.

INDIA PALE ALE » WET HOP INDIA PALE ALE

INDIA PALE ALE

Pale Ale changed the drinking habits of a nation, and then the world. Fitting for the style that gained a prefix indicating it was always destined to go far. As we detailed in our previous book, *Craft Beer for the People*, breweries in the UK had been shipping their wares to the Indian subcontinent for decades prior to the adoption of the IPA moniker – and most of the troops stationed there drank what they did back home; porter. But you can't keep a good pale beer down.

What was once the drink of the officer classes of the Raj became white-hot at home and pushed porter even further from the stage. What stout started, India Pale Ale resoundingly finished. They eventually became lighter in body over time and gradually morphed into the style that is also an adjective describing its appeal – Bitter – but true India Pale Ale was never lost to anybody. It has always been with us.

The reason why isn't rocket science – it's the main reason why people drink beer. The flavor. India Pale Ale has come to dominate the conversations of a small section of the beer world, and is set to take on the rest, because it is just so damned tasty. With its instantly recognizable abbreviation leaping off the label of dozens of cans and bottles every time you visit a bottle shop, you know what you are going to get.

What you'll get, pretty much, is hops. People love hops. From home-brewers who start careers with small packets of dry Cascade to tourists arriving in an unfamiliar city brewpub before asking the question we've all overheard – "What IPAs do you have on tap?" – the pithy, the resinous, the citrus, and the tropical are flavor families beer drinkers just can't get enough of. IPA is here to stay.

SIERRA NEVADA NORTHERN HEMISPHERE HARVEST WET HOP IPA (CHICO, CALIFORNIA)

There are more ways to riff on the original. Our path from IPA took many different routes and resulted in a multitude of flavors and aromas..., but there was a thread of commonality linking them all together. Not the hops used, or how many, but how those hops were processed before they even arrived at the brewery. The vast majority of the beers you drink will be brewed with dried hops.

After harvesting, commercial hops are kilned much like barley is roasted into malt. This sets their aromas, gives them a longer shelf life and enables the hop grower to ship them around the world to where they are needed before they go moldy. But there is an alternative. What if, instead of the hops losing 70 percent of their moisture to the hop kiln, you could transfer that into a beer? Well, you could create such a beer, but you'd have to move quickly.

Wet-hop IPAs are the pinnacle of logistics. Picked, packed, and dispatched hops go south fast, whether the brewery receiving them is north, south, east, or west of the fields. That high moisture content works against everyone involved, which is a reason why many of the first examples appeared from breweries within a stone's throw of the farm they selected their hops from.

Wet hops are used in exactly the same way as their dried stablemates, even for dry-hopping – or wet-dry-hopping. The unkilned cones are powerful nuggets of aroma, with massive reserves of hop oil unfettered by anything other than a speedy journey. The forerunners of this style, as with many others, are Sierra Nevada. Their Harvest Ale features hops picked 24 hours previously; it is intense, herbal, bitter, and will change the way you look at IPA. Again.

SEE ALSO:

FARMAGEDDON WET HOP IPA (NORTHERN IRELAND)

Hops from the brewery's own bines are harvested 15 minutes before being deployed as the final addition to this sappy, resinous, and botanical India Pale Ale.

RHINEGEIST WET HOP (CINCINNATI, OHIO)

Fresh Chinook and Cascade from Knightstown, Indiana, make this Cincy concern's WHIPA a piledriver of grapefruit and lemon-rind bitterness, but you'll have to go to their taproom for it.

VICTORY HARVEST ALE (DOWNINGTON, PENNSYLVANIA)

Another lively offering with the alpha acids from the just-picked hops coming over as a grassy, peppery bouquet thanks to six tons of hops that were trucked across country to their brewery.

WHEAT BEER » FRUITED WHEAT BEER

WHEAT BEER

As with stout, pale ale, and lager, when you think of wheat beer, what is typically noted to be a single type of beer is actually a family of many. Brewing with wheat has been a thing for nearly as long as wheat has existed, whether cultivated or foraged from naturally occurring thickets. The history of brewing features wheat as a wingman to barley since the year zero, and shows no sign of slowing down – even if, over that time, it has returned from the brink a few times.

Stop us if you've heard this before, but when looking at the forebears of the wheat-beer family, the best place to start is either Germany or Belgium. Or, preferably, both. Weissbier and witbier are as formative as you can get, having inspired breweries to reach for something other than malted barley for generations. And yet, generations ago (or more recently in terms of witbier), they very nearly left us forever. And the reason? Both times: lager.

In the mid-1800s, Bavarian weissbier was losing in the popularity stakes to lagers, as the rights to brew the wheat beer of the day were held only by breweries owned by regal dukes. Fast-forward 100 years and the white beers of Belgium were facing similar competition from the lager brewers of the day. The opaque, floral offerings from the past suddenly became antiquated and almost remained there.

Belgium has a milkman named Pierre Celis to thank. He pulled witbier from the doldrums with Hoegaarden, around 90 years after a brewer to the dukes by the name of Georg Schneider did the same with Bavarian wheat beer. Both Schneider and Celis plugged away, doggedly dragging their respective wheat-laden throwbacks into a modern age. Today, thanks to their perseverance, Schneider Weisse and Hoegaarden are household names (in beer-drinking houses).

21ST AMENDMENT BREWERY HELL OR HIGH WATERMELON (SAN FRANCISCO, CALIFORNIA)

Today, just as Celis's De Kluis and the Schneider breweries pulled wheat beer up by the bootstraps, brewers are continuing their work and bringing the glory of wheat to a new audience. But this time, all bets are off.

Originally, the Bavarian wheat beers allowed only wheat to be used in addition to barley, hops, malt, and yeast, whereas the traditional witbiers of Belgium had wiggle room thanks to the addition of orange peel, coriander, and other spices. It is that playbook the brewers of today are using. Few beer families, or even styles, are more accommodating to unusual ingredients than wheat beer. Their refreshing nature and classic estery profile from the yeast, which results in flavor characteristics such as clove, banana, herbal, bubblegum, and floral notes, are lifted and augmented by all manner of additions. But one alone adds up to more, and that is fruit.

From the bowl on your kitchen counter to the beer glass in your hand, fruited wheat beers are the new normal when you head to the bottle shop. Replacing orange citrus and fragrant coriander seeds, we now have every tropical, berry, and orchard fruit you can think of. If it'll attract a wasp at a picnic, it's fair game. Fruit is hugely varied, adding flavor, aroma, sweetness or tartness, as well as color to the beer. There's life in the ol' wit just yet.

For the ultimate refreshing summer hit in wheat-beer form, San Francisco is the place to go: 21st Amendment have cornered the style thanks to the hefty ground-dweller *Citrullus lanatus*. The watermelon, which is actually a berry (because of course it is) gives a delicate sweetness and floral fragrance to their 4.9% ABV American Wheat Beer. Crisp, dry, fruity, and served with a wedge of the berry fruit itself, this is summer beer *par excellence*.

SEE ALSO:

BREW BY NUMBERS 07 WITBIER RASPBERRY & HIBISCUS (ENGLAND)

A perfect kick of sharpness to an aromatic wheat beer is delivered by raspberry, which apparently isn't a berry at all, even if BBNo's example is near perfect.

GREEN FLASH PASSION FRUIT KICKER (SAN DIEGO, CALIFORNIA)

Adding sweetness, richness, floral notes, and body to a wheat beer is the aim of every fruit addition, and the perfect fruit for that is the passion fruit, which actually *is* a berry.

ANCHOR MANGO WHEAT (SAN FRANCISCO, CALIFORNIA)

Californians have grown mangoes since the 1850s and Anchor's year-round 4.5%-er features this sweet, nectar-like juicy fruit (it is a fruit!) in perfect harmony with the wheat malt.

GUEUZE » KETTLE SOUR

GUEUZE

Created in breweries that resemble museums sprung to life (even today), Gueuze is about as out -here as beer gets. The "toughest first-taste in beer™" is the application of careful husbandry over spontaneous mystery. As we detailed on page 31, windows left open overnight allow wild, wind-borne yeast to settle on and ferment wheat beer brewed by those around Brussels in Belgium. This lambic is then blended by those with palates to rival any other into gueuze.

Modern Beer Geeks espouse the virtues of lambic like few other styles, and yet the majority of lambic they are lucky enough to try will be, in actual fact, gueuze. Straight-up lambic is unblended and usually served young, flat, and sour from the breweries that craft it, sometimes even from tabletop pitchers. For that, you have to go to the source, in what will clearly be a day to remember – but for all other days, we have the bottled gueuze to turn to.

And what a direction to take. Gueuze is magical. The name derives from the Flemish word equating to "geyser," so if one is mishandled and then opened near you, look out. These beers are more carbonated and more complex than lambic; they are tart, sour, funky, and all manner of other adjectives. It was the advent of machine-fabricated glass bottles that saw gueuze production leap forward, but if you dig hand-blown glassware, you'd probably love gueuze.

The blenders (*gueuzestekers*) often existed solely to blend the brewer's work into their predetermined results. They were the quartermasters of the lambic, guardians of the style. To have the trust placed in you to know every inch of the wooden foudres in your warehouse, and how each was progressing at any one time, is a weight that few can appreciate. Gueuze is a relic, but one where everything about it leaves you more astonished.

BUXTON V OMNIPOLLO LEMON MERINGUE ICE CREAM PIE (ENGLAND/ SWEDEN)

Until this arrives at your table. What even is this? Reading off a beer menu would make you think you'd picked up the dessert options by mistake. But the process harnessed by the new titans of alternative souring – Sweden's Omnipollo – is a way to create all of the flavor and charm of the gueuze but without resorting to wooden-panelled breweries where the spiders outnumber the humans. Today, we also have the kettle sour.

It's a kind of quick-sour method, really, although that term is slightly too pejorative. Instead of fermenting spontaneously and/or re-fermenting in the bottle for months in the company of cobwebs, these beers are soured in the brew kettle (or, more typically, in a similarly stainless-steel fermentation vessel). Bacteria is added to the brew and results are seen within weeks, rather than months.

Before the yeast is pitched, lactobacillus is added, giving it a head start to have at the sugars and tip the balance of the beer into the low-3s on the pH scale. Even as the yeast finishes up, the beer remains sour and tart. Brewers can either culture up a Lacto strain or really rip up the rulebook and add something like natural yogurt. This is a balancing act, but if you know what you're doing, it's a pathway to amazing sours.

Chief among those who know what they are doing are Omnipollo. For their collaboration with Buxton they also added lactose to give a sweetness to the sour and to counter it perfectly. To complete the depiction of the dessert, they then conditioned the beer on lemon juice. And it really does taste like lemon meringue pie. This is beer that makes you ask questions with a smile on your face.

SEE ALSO:

GAMMA PURP (DENMARK)

The sharpest ingredient in the culinary arsenal is surely rhubarb. Denmark's Gamma have combined it with blackberry in this dry, sour, lactic refresher.

THORNBRIDGE TART (ENGLAND)

Its name leaves you in no doubt: this Bakewell delicacy was brewed with Wild Beer Co and bitterly sour lemon from start to finish.

MAIN STREET FRUIT BOMB (CANADA)

This pale, tart sour takes its own separate direction thanks to a hop-bursting technique that thunders the beer along with tropical and grapefruit flavors.

RAUCHBIER » SMOKED BEER

RAUCHBIER

Having confidently stated that Gueuze was the toughest first-taste in beer™, we now move on to the walk in the park that is the smoked beers. These "hot dogs in a blender" moments give you campfire flavor in a bottle, yet although the first take is a tricky one, with each sip you can be confident you're drinking something truly historic. Drying and roasting barley over open fires was how it all started; at one time, smoked beer was less a quirk than the norm.

Over time, maltsters developed ways of making the smoke from the fire bypass the malt bed, and it was left to a lone area of southern Germany to keep things as they were. The heartland of rauchbier ("smoked beer") – then as now – is Franconia, and at its epicenter is the place where smoke has become a way of life, Bamberg. This picture-postcard town of arched bridges and medieval and baroque buildings should be on any Beer Geek's bucket list.

Here, breweries like the world-beating Schlenkerla and Spezial use the power of beech-smoked barley malt to get the characteristic dusky tang into their beers. Franconians have huge native beechwood forests on their doorstep, and it takes only the smallest leap of the imagination to think back to their home-brewers of old felling trees to feed the fires that turned barley kernels into sweet, smoky rauchmalt, as a handful still do today.

Other wood can be used to apply the aroma and flavor of smoke to barley as it is kilned, as we shall see in a moment, but the beech-smoked beers of Bavaria are as unique today as they were commonplace back in the day. Bamberg is a UNESCO World Heritage Site for its architecture, but its rauchbier should also gain a similar listing. These beers are a taste that, once acquired, will never leave you.

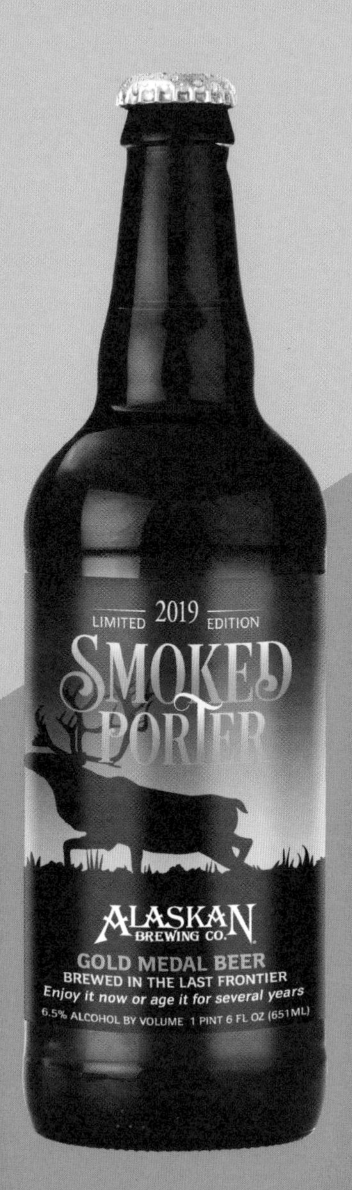

ALASKAN BREWING CO. SMOKED PORTER (JUNEAU, ALASKA)

Somewhere it can also be tricky to leave is the city of Juneau, Alaska. Arrivals or departures are made by air or water, as it is the only state capital with no roads in or out. Juneau is a frontier city in every sense of the word. Back in the winter of 1986, Marcy and Geoff Larson founded a brewery there, the first since the Prohibition era, and it quickly gained a hugely passionate local following (as you would kind of imagine).

As befits a brewery that focuses on the uniqueness of its isolation rather than the logistics of its situation, Alaskan Brewing Co. naturally embraced the art of smoke beer. The brewery sought to make a true Alaskan product from the get-go, using spruce tips, aging barley wines in gold-rush tunnels – and so crafting their own version of the German rauchbier. It was a moment of genius.

They now create their own rauchmalt, smoking it over local alder instead of beech, and brew the beer with glacial water. This is a craft brewing in a nutshell – Marcy, Jeff, and the team took inspiration from at home and overseas and married them together in a unique product born of the resources around them. Smoked beers started local, as the only available option to brewers. Alaskan Smoked Porter is a new personification of that.

It was first released in 1988, and if you can get your hands on a bottle it will be well worth the effort. There is a lot of dark, bitter chocolate and cocoa from the unsmoked malt, but the campfire character arrives on the finish – it is never dominant, though, or overpowering. This is campfire appreciated at a distance, not a sudden wind-changing faceful of smoke. Alaskan Smoked Porter is a perfectly balanced beer that keeps on giving, the more of it you enjoy.

SEE ALSO:

STONE SMOKED PORTER (ESCONDIDO, CALIFORNIA)

California's Stone Brewing first released their smooth smoked porter back in 1996, and the power behind their throne comes from dark, heavily flavored peat-smoked malt.

VIVEN PORTER (BELGIUM)

Sometimes referred to as their Smoked Porter, whether so-named or not, Viven's example has masses of dark, stewed fruit alongside the smoky aroma and flavor.

TOP OUT SMOKED PORTER (SCOTLAND)

This Edinburgh example is brewed with both beech-smoked and peat-smoked malt and is the smokiest of the lot – if you like the peated whiskies of BrewDog's home nation, look for this.

PALE ALE » GLUTEN-FREE PALE ALE

PALE ALE

Way back before the fires of industry took hold, there's a temptation to think all beer was dark, weighty, and taken sullenly in earthenware mugs in ramshackle taverns. The taverns may have been that way, but the beer? Not always. Pale Ale seems like a modern addition to the brewer's locker thanks to American beer, golden ales from the Midlands of England, and other bottled staples we now find everywhere.

But those beers are just that – *additions*. Historians such as the great Martyn Cornell have been explaining, patiently, for years that right back to ancient times brewers have used malt that had been dried by being laid out in the sun. Before the development of coal-fired kilns, if you didn't want smoke from wood affecting your drying barley, leaving the sun to do the work was one of the only options.

As anyone who has left an old baseball cap in the back of the car knows, long exposure to the sun bleaches color from things, and sun-dried malt would, as Martyn argues, undoubtedly have led to pale beer. Millennia before India Pale Ale, Sumerian Pale Ale was a thing (although, sadly, it was over 3,000 years before it could be checked in on Untappd). Once coal – or its derivative, coke – came on the scene, pale malt became easy to obtain. Pale Ale was here to stay.

Darker beers gained a following in the 18th century, as we've seen from the porter-into-stout conundrum. Not to be outdone, their alternatives in the pale ale family also had their own naming confusion. Many breweries created different pale ales – as they do today – and some drinkers began calling them "bitters" to differentiate them from other beers they could buy, such as pale milds. Whatever you call it, pale ale has been with us since beer began.

BREWDOG VAGABOND GLUTEN-FREE PALE ALE (SCOTLAND)

Pale malt doesn't just form the backbone for pale ales; dark beers have more of it in their make-up than darker malts. For instance, our 16.5% Imperial Stout Tokyo is blacker than the heart of your high school principal, but its ratio of malt is 80:20 pale to dark. This powerful stout takes the vast majority of its power from extra pale malt. But as all beers major in malt, whatever their shade, they will also contain something many people cannot stomach: gluten.

These proteins, which are found in most of the grains that become malt, can cause major problems for those with coeliac disease and other gastric intolerances. At BrewDog, we believe in making beer as accessible as possible and so developing a gluten-free pale ale was high on our agenda. Back in the summer of 2015, we launched Vagabond into the world, and it has since become one of our best-selling beers.

With so much malt here, how do brewers work around it? There are a couple of ways – such as mashing in with grains that are non-glutinous (sorghum, buckwheat, oats, millet, to name a few); or you can adopt a touch of technology and use an enzyme added at the start of fermentation. This acts on the intolerance-causing proteins and breaks them down before the beer has been packaged, leaving it well within the gluten-free threshold.

The advances in gluten-free options have been enormous over the last few years (in all sectors, not just brewing), and with Vagabond we wanted to show, once and for all, that gluten-free need not equate to taste-free. It is citrus-and pine-resin-forward from the Amarillo and Centennial hops, with a touch of tropical fruit on the palate at the same time. The light caramel malt base and the resinous bitterness work enormously well together.

SEE ALSO:

BURNT MILL STEEL CUT (ENGLAND)

Brewed using a blend of oats, buckwheat, maize, and sorghum, this beer is smooth and balanced, still maintaining their trademark stone-fruit/tropical-fruit flavor profile.

MIKKELLER PETER, PALE AND MARY (GLUTEN-FREE) (DENMARK)

Mikkeller's "Folk Pale Ale" has gone gluten-free and sings with grapefruit zest from start to finish. Well worth seeking out.

TWØBAYS PALE ALE (AUSTRALIA)

Founded by a coeliac sufferer, this Victoria-based brewery has an entirely gluten-free line-up. Their flagship pale ale is brewed with millet, rice, and buckwheat.

BRITISH BARLEY WINE » AMERICAN BARLEY WINE

BRITISH BARLEY WINE

Being near-neighbors, Britain and France share a history that is long and involved, and which frequently turned bloody. So much so, there were not one but two multi-generational conflicts deemed large enough to be termed "The Hundred Years' War." And it is the second of these that may well have resulted in the birth of one of the broadest beer styles of them all – the barley wine.

These were beers born of necessity. Strong ales brewed by the houses of England, they were designed to be aged in barrels in the cellars and drunk as "old" ales – matured like fine wines. When actual, real conflict interrupted the supply of French wine, you had to fall back on what you could brew yourselves. What these beers were like is anyone's guess – but they must have been full-bodied, rich, and hearty.

Fast-forward and, as relations with France improved, the Midlands town of Burton upon Trent took up the mantle. Strong Burton Ales like Bass's No. 1 or Ind Coope's No. 1 brought these types of beers to a new market. Over time, that market was enticed with beers named "barley wine" as well as "old ale," "Burton Ale," "strong ale," and any one of a number of different newly emerging brand names.

When poured, these beers arrived in every possible shade from golden to dark – but they were all strong. Barley wine essentially became a catch-all term for any uplifting beer that wasn't an imperial stout, but they shared a common thread in their potency. If you'd been able to order one at any point through its long, evolving history, it would have definitely required the immediate solace of the nearest armchair.

KEES AMERICAN BARLEY WINE (NETHERLANDS)

It's tempting to read that and think barley wines are long gone, drunk only by men in tri-corner hats smoking curving clay pipes, but breweries in the traditional heartlands of barley wine – Burton, London, Scotland – still brew them to this day. Classified as Old Ale or English Barley Wine, as befits the keepers of the style guidelines, these full-bodied, warming beers remain to be discovered. And they also remain influential.

As a legion of Americans began firing up their brewkits in the 1970s and '80s, it was to Europe that many of these pioneers looked for inspiration. It's not difficult to imagine them reading about the high-gravity, heavily flavored barley wines of old and wanting to re-create them. It was Anchor Brewing Co, back in 1975, who released the first modern-era American barleywine-style ale: Old Foghorn.

We have seen elsewhere that one of the underlying themes of American craft-brewery-created "takes" on the historic styles of other countries is to be as faithful as possible but to increase the hop dosage as they see (or saw) fit. American barley wines are big and bold, but also citrusy and/or resinous from the hop load. It's a fine line between these beers and Double IPAs, but it's a fantastic line for the drinker to tread.

In fact, you can age a Double IPA, either deliberately or by being forgetful, and experience just how close it leans to something like Old Foghorn. But these lines needn't be blurred by the drinker alone; we now live in a world where European breweries are putting their stamp on American barley wine, in the ultimate act of return to sender. Check out Brouwerij Kees' 11.5% ABV version, which thunders with dried fruits, caramel, and a sherry-like quality.

SEE ALSO:

LEFT HANDED GIANT LIFE WITHOUT OXYGEN (ENGLAND)

Caramel and spice at the start, leading into a long boozy finish from Bristol's finest. There's also more than a hint of tropical fruit in that long finish.

GARAGE PROJECT HELLBENDER (NEW ZEALAND)

A ton of malt in every batch, this Kiwi barley wine is all about the dark stonefruit – dates, in particular. Also sweet, sugared figs at every turn.

SIERRA NEVADA BIGFOOT (CHICO, CALIFORNIA)

Buzz Aldrin to Anchor's Neil Armstrong, the second American barley wine-style ale arrived in 1983 and is, as Sierra Nevada rightfully claim, a cult classic.

BALTIC PORTER » BARREL-AGED IMPERIAL STOUT

BALTIC PORTER

If you think that breweries exporting their wares around the world is a modern phenomenon, then let Catherine the Great politely tell you otherwise. The Empress of all Russia, who overthrew her own husband in a coup to take power, was partial to a drop or two of London-brewed porter. Having her hands full being the longest-ruling female leader in Russian history meant that the beer had to come to her, not the other way around.

As a result, England exported porter and strong porter (in other words stout) from the docks of its eastern coastline across the North Sea and into Russia, Poland, and other countries there that lined the Baltic. Burton Ale also went in the same direction, but it was the darker beers that caught on, and eventually became synonymous with the name of the marginal sea across which the 19th-century wooden ships plied their trade.

As you can imagine, brewers in that part of the world didn't take this lying down – or rather, they saw an opportunity to get in with a popular beer of the time – and began to create their own versions. The trouble was that breweries in the Baltic region relied on "bottom-fermenting" yeasts suited to lagers, rather than the "top-fermenting" ale yeasts of England. So their own versions were far from the real McCoy. Exported Baltic porter was here to stay.

Except, it wasn't. Catherine's reign ended and tastes changed. Western Europe's relationship with the Russian Empire changed, cooled, and then became outright hostile with the outbreak of the Crimean War. Beer no longer moved across the Baltic. But its legacy remained, on both sides. Around the Baltic, breweries continued with their cold-conditioned strong porters, and in Britain the public gained a taste for a beer that had always left them for other shores.

HARVIESTOUN OLA DUBH 12 (SCOTLAND)

With the home nation now enjoying the strong, ale-yeast-powered "imperial" porters that had previously been added to barrels and shipped overseas, Russia's loss was Britain's gain. As we have seen, strong porters were typically termed "stout" in the alehouses of the time, and so the term "imperial stout" duly arrived. Once the conflict in the Crimea had passed, "Russian stout" appeared in the parlance as well.

Fast-forward to the present day and in the dark-beer pantheon, "Russian stout" or "Russian imperial stout" is heavily, heavily outweighed by the snappier "imperial stout" in modern naming conventions. A few exist that give a shout-out to its history (North Coast's Old Rasputin and Thornbridge's Saint Petersburg are two excellent examples), but if you rock up at a taproom looking for a strong, dark beer chances are it will be "imperial stout" your eyes alight on.

These days, and actually for longer than you may realize, brewers are adding another prefix to it – and it's one that even Catherine herself might approve of. Brewers are returning the beer whence it came by adding their weighty stouts to wood. "Barrel-aged imperial stout" is with us, and is one of the most rewarding beers you can experience. he actions of time, casks, and added ingredients (if deployed) has taken imperial stout to the next level.

In the heart of Scotland, one brewery has been doing this for longer than most. In 2007, Harviestoun became the first brewery to formally partner with a distillery and introduced their Old Engine Oil stout into vintage Highland Park whiskey casks. Ola Dubh 12 arrived on the scene (*ola dubh* meaning "black oil"). Coffee, vanilla, dark chocolate, boozy truffles, and more await. The 16 and 18 are amazing; the Ola Dubh 30, if you can find it, is life-affirming.

SEE ALSO:

DESCHUTES THE ABYSS (BEND, OREGON)

Deep and dark, as per the feature it is named after, Deschutes' finest contains six malts, four hops, molasses, vanilla, liquorice, and cherry bark, and is simply spellbinding.

EVIL TWIN EVEN MORE JESUS BOURBON MAPLE SYRUP BARREL-AGED (DENMARK/USA)

Brewed at Westbrook Brewing this beer takes longer to say than pour, but a huge stout aged in maple syrup barrels? Yes please.

FYNE ALES BOURBON BARREL-AGED MILLS & HILLS (SCOTLAND)

Brewed in collaboration with De Molen. Scotland's Fyne Ales added this beer to bourbon barrels for a massive hit of sweet vanilla and coconut majesty.

GRUIT » PUMPKIN ALE

GRUIT

These days, it's a struggle to find a commercial beer that doesn't contain hops. The mini cones of magic have been imparting their aroma, flavor, and preservative character to beer since medieval times. In Europe, the trading towns of northern Germany were exporting hopped beer to other parts of the continent from around the turn of the 15th century. Hops have hung around for a while. But so, too, have herbs.

Whether hops were available to you or not, an alternative was to add botanical plants into your brew to give it flavor. But this came at a price – literally; these flavorings were controlled by the state, regulated and taxed to gain an income for the coffers of The Man. Churches, councils, and merchant families all became rich on the back of a mixed bag of herbs used to make ale ("ale" signifying it was unhopped, and therefore different to "beer").

The catch-all term for this collection of plants was *gruit* (in Holland) or *grute* (in Germany). Commonly used botanicals included sweet gale (or bog myrtle), yarrow, and wild rosemary. They added flavor, but also a kick. Beer historian Martyn Cornell has described accounts of too much yarrow inducing ringing in the ears, bog myrtle being used by Vikings to bring on pre-battle frenzied hallucinations, and of sweet willow leaves being made into a tea given to children to get rid of intestinal worms.

So were *gruit* beers a lottery? It must have depended on where you were and who was making them. It doesn't take a leap to imagine backstreet brewers concocting all manner of brews. Over time, this uncertainty joined with Puritanical doctrines banning frenzy-inducing ingredients (boo!). The Bavarian Purity Law arrived, the influence of the Catholic Church waned, and the trading ports were suddenly flooded with hopped beer. *Gruit* was consigned to history.

DOGFISH HEAD PUNKIN ALE (MILTON, DELAWARE)

Gruit may have disappeared, but only in its previous form. Beers do exist today that are brewed with these botanicals, released by those curious to see what they would have been like and drunk by people of the same mind; only minds are less altered these days and ringing ears are definitely a thing of the past. *Gruit* was a wide term; it included things like cinnamon, ginger, and other items we recognize from modern culinary use, if not exactly from brewing.

It's only a small jump from cinnamon to another flavor commonly associated with when the leaves turn and begin to drop from the trees. Pumpkin. We can safely say that these ground-growing gourds would not have found their way into traditional *gruit* as they are native only to North and Central America. No Viking warrior would have crushed a pumpkin with their teeth before a battle (unless it was following their transatlantic voyages).

In that land where the Vikings eventually rumbled around, pumpkin beers are a fact of life – head to any American bottle shop in the early fall and you won't have to look too far before you spy an orange label, or 20, on the shelf. In other parts of the world they are catching on, too; albeit the breweries have to get the pumpkins in and begin the brew while summer is still with them; such is the drawn-out nature of fermentation, conditioning, packaging, and delivery.

On the eastern seaboard, Dogfish Head are an institution when it comes to using unfamiliar ingredients, although having been founded in 1995, they are an institution full stop (or period). Among their weighty catalog is the seasonal Punkin Ale, brewed with pumpkin, brown sugar, and spices and released at the start of every September. It is warming, rich, and spicy with a kick of cinnamon and sweet dark sugar and a noticeable tang of pumpkin on the finish.

SEE ALSO:

CIGAR CITY GOOD GOURD (TAMPA, FLORIDA)

Pumpkin beers are primed for pun-based names, and Florida's Cigar City have a winner with their liquid monument to the "most noble of all the gourds."

ALLAGASH GHOULSCHIP (PORTLAND, MAINE)

In Maine, Allagash have worked miracles with this blended sour pumpkin ale, aged on oak after being left to ferment in their coolship overnight on October 31.

SOUTHERN TIER IMPERIAL PUMKING (LAKEWOOD, NEW YORK)

One of the most popular pumpkin beers is this firecracker from New York, which arrives at 8.6% ABV and appears in August every year.

OLD GOSE » NEW GOSE

OLD GOSE

When it comes time to sort the beer styles of the world into some kind of order of wackiness, the top spot is reserved for Gose. Sure, the spontaneously fermented beers of Belgium are up there – but they got their break thanks to the randomness of the breeze. Someone thought up the Gose. They sat down and pieced together a beer that is sour, salty, sharply effervescent, and herbal. All at once.

The style – which was also traditionally fermented spontaneously – is a true oddball; one of those quirks that if it was in the natural kingdom would end up with lizards that aren't lizards (see: the Tuatara). Gose is a fascinating mix of wheat beer combined with things you have within three feet of your kitchen stove.

Traditionally associated with the town of Goslar and the city of Leipzig, in Germany, Gose was brewed with wheat, barley, and sometimes oats, but also with coriander (typically crushed or ground seeds) and salt. Whether they used actual salt or salty water from the mineral-rich local supply, these beers all had that trademark tingle.

They were also bottled without corks or caps. After the yeast had gone to town in the wooden barrels where the beer was conditioned, the beer was transferred to long-necked bottles and left alone. As the still-active yeast slowly fermented away it would rise up the narrow neck and dry out, forming a plug in the bottle. Everything related to this beer style is fascinatingly strange.

NEW GOSE – MODERN TIMES FRUITLANDS (SAN DIEGO, CALIFORNIA)

Tastes change. People move on. Maybe Gose was too much of an oddball to survive in a world increasingly dominated by mass production; by conformity of flavor. Either way, Gose declined until it was brewed by a single brewery, which closed when the owner died in the 1960s, the secret of how to create the style reduced to entries in a notebook.

This dodo-like story has had a happy ending, however (unlike the dodos). Brewers in Germany found the notes and re-created the style. Gose blinked back into consciousness – and from there found the attention of the willing experimenters in the world of craft brewing. "A crazily flavored brew that was moments from extinction? Sign us up!"

OK, so Gose still isn't a household name, but modern brewers are having their curiosities piqued enough to give it a go. Crucially, there are both faithful renditions and versions with other complementary flavors added. The "let's take this and see what it brings to the party" mentality of craft beer has given Gose new legs, and new life.

Take Modern Times. San Diego is as far from Leipzig as you can get, but a brewery there has added a fifth element to the line-up of sour, salty, sharp, and herbal: fruit. Passion fruit and guava bring that other sensation into play – sweetness. And a big glug of tropical juice from Modern Times Fruitlands lifts and extends the umami baseline and the salty tang brilliantly. It's a masterpiece of brewing and a triumph of revivalist thinking.

SEE ALSO:

WESTBROOK GOSE (MOUNT PLEASANT, SOUTH CAROLINA)

The poster-child for craft-brewed authentically styled Gose. Right down the pipe and tasting amazing. This South Carolina brewery has nailed a Gose of which the citizens of Leipzig would be proud.

OMNIPOLLO BIANCA LASSI MANGO GOSE (SWEDEN)

The kings of the milkshake beer strike again. To elevate the Gose they channelled lassi, with *the* richest of fruits offsetting the salty sourness at every turn.

SIERRA NEVADA OTRA VEZ (CHICO, CALIFORNIA)

Think SN are all about the Pale Ale? Think again. The pride of California turned up the volume with this Gose also brewed with grapefruit and prickly pear. Yep, cactus beer.

BOCK » EISBOCK

BOCK

If, by chance, you should find yourself in Einbeck, take a quick wander around its historic center and look for a theme in the old, timber-framed buildings. This town in northern Germany is typified by houses with large, arched doorways that disappear deep into their interior. These aren't entrance halls, or stable yards; a different kind of beast was led in and out of these houses. Brewing kettles.

Back in the 15th and 16th centuries, every citizen of Einbeck had the right to brew beer – but it was a controlled privilege. None of the people were allowed to own brewing equipment, only the city council. Brewing became a kind of cooperative, with the leaders of the city owning the gear and the burghers doing the brewing. The equipment was carried into their house, used to produce a beer to a certified standard, and then carried out again to the next house.

As a result, the beer produced in Einbeck was regulated and consistent. With the city belonging to the Hanseatic League, it also became widely available and known for its strength and quality. The commonly told story is that, over time, the "Einbeck" became "ein Bock" to those who were asking for it, and since the German wording was a synonym for goat, that gave rise to the usual symbol of the bock being a goat.

As with lager, bock is a family rather than a thing, with different types being brewed depending on the time of year. Doppelbock, Weizenbock, and Maibock all accompany regular bock along the way, with another cousin making its presence felt (see opposite). All in the family are strong, malt-led and hugely rewarding, whether you are taking a quick pit-stop from that wander around Einbeck or simply heading to your sofa.

SCHNEIDER AVENTINUS EISBOCK (GERMANY)

Eis Eis Baby. Or more accurately, Eis Eis Maybe. The story of this very specific turbo-charged style is littered with references to happy accidents, fortune, and to the discovery that what should have been a disaster had actually created liquid perfection. When beer is frozen, the ice (or *eis*) can be left behind and the remaining liquid drained away, having become concentrated and fortified. Freezing beer makes it stronger. It also makes for a nice story.

In Bavaria (not Einbeck), brewers accidentally exposed bock to below-zero temperatures; beer left out overnight had frozen in the barrel, splitting the wood and wasting the lot. Even worse, it was a young brewery lad to blame; an apprentice, likely tasked with something menial that he either forgot or couldn't be bothered to do, got the blame for the frozen, splintered barrel now destined for the firepit. Until...

... Wait ... The center of the barrel was not frozen. What dark, stronger liquid was this? OMG, it tastes amazing!

We are paraphrasing here, but the oft-told story of a frozen barrel concentrating the bock inside into the darker, richer Eisbock has become part of brewing folklore. Were Eisbocks discovered by accident? Was this an accurate telling of their development or an embellishment of something minor that may or may not have happened? And would this count as craft or fortune? Does it matter?

Well, probably not. Every great product needs a good story, and Eisbock has powered a beery arms race. Freeze and Go has led to beers at 20%, 30%, 40% ABV, and above. Eisbock is a thing of beauty, an action of science over common sense. Take Schneider's thundering Aventinus version – at 12% ABV it has dark chocolate, dark stonefruit, banana, cloves, spiced nuts, port wine, and all manner of other flavors in there. This is a beer to make an evening of.

SEE ALSO:

MAMMOTH FIRE & EISBOCK (MAMMOTH LAKES, CALIFORNIA)

Brewed at altitude at Mammoth Lakes, California, this beer is put under more duress by being frozen, for a sweet, sherry-like raisin-pushing flavor high.

KUHNHENN RASPBERRY EISBOCK (WARREN, MICHIGAN)

In Michigan, this Eisbock is an annual release that runs between 10–15% ABV and is brewed with real raspberries, resulting in a jam-like, Black Forest gateau of a beer.

FREIGEIST STRAWBERRY EISBOCK FOREVER (GERMANY)

Speaking of jam brings us to this strawberry wheat Eisbock that is bittersweet, berry-heavy, and sticky and boozy on the (considerable, at 15% ABV) finish.

THE BARLEY FARMER

As we have said numerous times, without yeast there would be no beer. But without people like Will Hamilton, there would be nothing for the yeast to turn *into* beer. Like thousands of his fellow farmers around the world, Will works all hours growing the backbone of your pint – barley.

"What I do has a different feel to a nine-to-five job," he says. "It's a way of life. I'm never away from work. Well, unless I'm on holiday. When I walk out of my back door, I'm at work." You get the feeling any holidays he takes are short and spent with an eye on the phone in case something happens on the 2,500 acres he farms with his family just south of Coldingham in East Lothian, Scotland.

He ends our conversation by saying, "If you need anything else, you can catch me from five in the morning to ten at night, seven days a week."

If you look up Coldingham on an online map and click the satellite view, you'll see two things. First, the cluster of buildings and spidery roads are completely surrounded, on all sides, by fields neatly lined by ploughing. Second, if you zoom out a little more you'll see the immediate dark blue of the North Sea, a stone's throw from Will's farm.

"We farm in our own microclimate," he says, when asked how being so near the sea influences things. "There's not as much frost to affect the barley. The summer rarely has scorching days. We get a breeze from the sea; we get the *haar*, too." This unique-sounding word describes the summer sea fog, rolling into Scottish coastal areas as the warm air leaves the land and hits cool air over the cold water. What all this means is that his particular area of Scotland is ideal for growing barley, which he then sells on to maltings to start the process that will eventually lead to beer (and whiskey).

Will grows three kinds of the grain: winter brewing barley; heritage varieties, such as Golden Promise; and conventional spring barley, such as Laureate and LG Diablo. Two crops a year, plus the heritage varieties, might explain the comprehensive patchwork of fields around Coldingham. Around September every year, the farm drills barley into the earth from behind a tractor – and this "winter" crop remains in the ground for around ten months, being harvested in July (depending on the weather, more of which in a minute). Toward the end of March, they also drill the spring varieties, which are harvested toward the second half of August, having spent five months in the ground. These are typically higher-quality types of barley, grown on a shorter span through the warmer months and, as such, yielding less than the winter brewing barley crop. Sunshine builds yield, but the waving ears of barley can suffer on scorchingly hot days.

"Everything's about balance in farming," summarizes Will. "If you then get decent weather at harvest, it's usually OK. Sometimes over-ripening is a problem. A lot of wet/dry, wet/dry days leads to skinning, where it's hard to harvest without separating the grain from the husk." When this happens, the barley kernel won't germinate, so is useless to the maltster.

"Sometimes, the grain can pre-germinate in the field, so when it dries out, it kills the endosperm," he finishes, listing another potential way in which all of the patience and work on the fields can end up in barley being sent for animal feed, and a much lower price. Does it not keep him up at night, with all the potential pitfalls? "No, not really. I do think about rain events but I only really get uptight about quality."

Farming is surely the profession above all others where you are in control of your own destiny. Not complete control, as the weather can interject whenever it wants, but the crop is there, staring back at you every day of the week until you decide it's time to fire up the heavy machinery and start the harvest. You need dedication – and also, clearly, you need to truly love what you are doing in order to make a go of it.

Will uses the word love a lot; he's obviously in his element. The idea of growing things, the positioning of yourself and what you do in a food chain that leads elsewhere – that's possibly the reason he got involved with heritage barley varieties. These traditional, historic grains have been re-sown and given new life. He grows Golden Promise, a variety popular in the UK in the 1970s and '80s, particularly with Scottish distilleries.

"I really enjoy the connection you get with something like Golden Promise. It's a buzz really, seeing how their price and yield has changed over the years." The demand for it reduced dramatically in the 1990s but has returned, and Will is keen to help retain the link to older varieties – especially with Golden Promise having had its genetic make-up mapped, better to understand disease resistance.

When it comes to these heritage barleys, there is another way in which they can be used to further our understanding – the impact of climate change. As they have been farmed for decades, yields, growth, height, length to maturity, and other diagnostic characteristics have been recorded for some time. Will has grown heritage barley for 30 years – and has noticed a difference.

"Having grown heritage, I can compare back on harvest dates – it is usually August to September, but now I'm harvesting it fourteen days earlier. This isn't down to the variety. That hasn't changed. It's the climate. The barley is ripening earlier. In my business, it's actually beneficial as we can harvest drier and run longer combining days. But at the same time we are also seeing more extreme rainfall events."

Few people are more in tune with their surroundings than those who farm the land they own. And when it comes to the beer you drink, every time you open a bottle or a can you have the round-the-clock commitment of people like Will to thank (and the yeasts, don't forget them).

"WHEN I WALK OUT OF MY BACK DOOR, I'M AT WORK."

BRASSERIE DUPONT (TOURPES, WALLONIA, BELGIUM)

DRINK NOW:

SAISON DUPONT

ABV: 6.5%
STYLE: SAISON
COUNTRY: BELGIUM
AVAILABLE: YEAR-ROUND

A stone-cold Belgian classic. This is a beer to buy and drink as soon as you get home. On the bus, if you can. Brewed by the Dupont family and their descendants since 1844, it can be aged, but the flavors are so immediate and rewarding that not reaching for the bottle opener the moment you bluster into the kitchen becomes an insurmountable task. Whereisit… whereisit… whereisit…

And when you have that opener, what a reward. Bitter, refreshing, and enormously addictive, beers such as this one, that define an entire style, deserve to be investigated and enjoyed the moment you get home from the bottle shop, or standing in your kitchen surrounded by polystyrene packaging chips as the courier is still walking away from your front door.

This is a beer that was traditionally brewed during the winter and then served to the "*saisoniers*" who were working on the fields. These itinerant laborers moved with the harvest, and as they toiled under the hot sun they needed refreshment to keep them going. This style, fermented at the brewery and then again in the barrel (or, these days, in the bottle), is light and alive with citrus and spice. Drink it now.

No country on earth does beer history like Belgium. A few dozen miles from its undulating border with France lies the small village of Tourpes, and a farm that dates back to 1759 (the year George Washington got married). This is the home of what was the Rimaux-Derrider farm-brewery, a concern that specialized in brewing and storing seasonal beers for farmworkers – a.k.a. saisons. In 1920, when a local man learned his son was about to emigrate to Canada to start a farm, he bought the farm-brewery to keep him closer to home – and it worked. The son was Louis Dupont and his reimagined saison changed everything for him, as well as for generations of Belgian beer drinkers and for those who crave the vibrancy of its flavor.

DRINK NEXT:

SAISON DUPONT AVEC LES BONS VOEUX

ABV: 9.5%
STYLE: SAISON
COUNTRY: BELGIUM
AVAILABLE: DECEMBER–FEBRUARY

Saison, as a style and thanks to the history of how it was made and intended to be used, is something that can be aged to gain greater depth and power of flavor. And you don't need to go far from Dupont's most popular offering to find a perfect example – although you will need to wait until a certain time of year (originally, you would also have had to be in the good books of those in the brewing family).

Saison Dupont *"Avec Les Bons Voeux"* ("with best wishes") is traditionally created by the brewery as a New-Year gift to their most loyal of customers. It is a celebration; a calling card of all that the brasserie are about and wish to pass on to those who enjoy their art. It can be purchased, though, so if you don't live within cycling distance of Tourpes, no worries. And when you get your hands on one, why rush it?

Weighing in at just under double-digit ABV, this beer is crying out to be aged. The Belgian yeast in the bottle will continue to support the flavors as they round out over time, giving a deeper, less prickly, and supremely balanced result. It is rich, velvety smooth, and would make an ideal New Year's Eve beer – this year or next. Some beers are meant to be aged – cellar this on our advice and with Dupont's best wishes.

BELL'S BREWERY
(KALAMAZOO, MICHIGAN)

DRINK NOW:
BELL'S PORTER

ABV: 5.6%
STYLE: ROBUST PORTER
COUNTRY: USA
AVAILABLE: YEAR-ROUND

When we run down a list of beer styles that taste best when immediately acquired, they aren't all hop-led pale ales. Dark beers have an immediacy about them too, particularly if the trip to get them was made in poor weather, or at the time of year when the sun dips beneath the horizon before you've even decided what you'll be having for dinner. Sometimes, you just need a welcoming hug from a porter.

When it comes to hugs, robust is best. Robust porters are amplified by a kick of alcohol that firms through a finish already made rewarding thanks to dark, roasted malts (typically black patent malt) in the recipe. These give a bitterness to the beer, and make the style something akin to a malt-led cousin of the Black IPAs that lead with hop-derived bitterness.

Bell's Porter is a fantastic beer and, as such, one that really doesn't need any time on it. Whether the day is grey, dark, or blazingly sunny, uncap this one as soon as it has reached a temperature you like your dark beers at. Chocolate and coffee greet you in unison before the rich, warming finish puts the best kind of emphasis on whatever kind of day you are having. Hugs all round.

Every brewery knows the exact date when their first beer went on sale. It will have been recorded in a ledger or diary, or exist on the bottommost sheet in a pile of papers on a desk somewhere. In modern times, there will be a date-stamped social media post online. For Larry Bell of Bell's Brewery in Michigan, the date his first beer went on sale was September 19, 1985. His brewery was founded on three things: an interest in fermentation, a 15-gallon (68-liter) soup pot, and a change in the law making home-brewing legal. That keenness for the action of yeast was developed by Larry working at a bakery in Kalamazoo, and when he took advantage of Jimmy Carter relaxing the ban on home-brewing, he left and opened a home-brew supply store. With that old soup pot as his brewing system, one of America's best breweries was born, on that exact date in mid-September, 1985.

DRINK NEXT:

BELL'S EXPEDITION STOUT

ABV: 10.5%
STYLE: RUSSIAN IMPERIAL STOUT
COUNTRY: USA
AVAILABLE: SEPTEMBER

On the other hand, we have this monster. If Bell's Porter is a robust hug, Expedition Stout is a bear hug. The brewery not only suggests this beer is ideal for aging, they enter aged versions of it in the industry's leading competition, the Great American Beer Festival. The 2011 event awarded a Gold Medal to Bell's for an Expedition that was, at the time, 16 years old. You can't argue with that.

In fact, the official shelf life for this beer is listed by the brewery as "unlimited," so if you can find it as the leaves start to turn, buy as many as you can and stash them somewhere safe for the foreseeable future. This is the perfect beer for a vertical tasting, or to bring out every few years and see if this one has developed different flavors from the last.

Expedition (it's usually known by that single name, Madonna-like) is a massive beer in every respect. Huge, boozy aromas of dried fruit, stewed fruit, dark, bitter chocolate, and roasted coffee beans abound, whatever vintage you have in your shaking hands. With some age on it, these flavors will have merged into one glorious whole that is about a perfect a beer experience as you can get.

LITTLE EARTH PROJECT (SUDBURY, SUFFOLK, ENGLAND)

DRINK NOW: LITTLE EARTH PROJECT ORGANIC HARVEST SAISON

ABV: VARIES
STYLE: SAISON
COUNTRY: ENGLAND
AVAILABLE: YEAR-ROUND

Every great saison should evoke a pastoral way of life. Fields and orchards have been in the DNA of the style since the original versions gave afternoon refreshment (and not a little sustenance) to traveling workers and pickers. They are made to be drunk immediately (as we mentioned at length on page 110 with the category-defining Saison Dupont).

Crucially, back in the day, they were also different each year. Conditions, ingredients, and techniques varied as another winter rolled around – of course, the beer that appeared the following summer would be unique. It was the skill of the saison brewer to work with what they had and construct a beer that everyone who drank it would appreciate.

That spirit comes through in this amazing annual release from Little Earth Project that embraces oak aging and wild yeast. Dry, spicy, and with more than a hint of cidery orchard notes, each vintage of this beer will be different and, as such, will be like stepping back in time. What could be more immediate and pressing to discover than that?

Countryside breweries specializing in saisons brewed in touch with the local landscape aren't just to be discovered in the low countries of mainland Europe. A mere 60 miles from Greater London is the market town of Sudbury; nearby, a dark-timbered stable building – the home of Little Earth Project. Founded in 2016 by Tom Norton, the historic and farmhouse styles that emanate from within are made with the land in mind – ingredients are either seasonally foraged or organically farmed. Their hops and malt are grown three miles away and their wild yeast originates from apple skins used by Tom's family to make cider. This is throwback brewing for the modern age. Little Earth Project are as sustainable – and as intriguing – as they come.

DRINK NEXT:

LITTLE EARTH PROJECT FOLLY ROAD

ABV: VARIES
STYLE: SAISON
COUNTRY: ENGLAND
AVAILABLE: YEAR-ROUND

If anything has become apparent in the pages of this book, it is that even the broadest of beer styles have a huge amount of variation within them. The official list curated by the Beer Judge Certification Program increases in size with every release, but somehow you still find yourself reading a bottle label and thinking, "well, where does this one fit in?"

For an example, we have Folly Road. It is another annual-release saison, also brewed by Little Earth Project and featuring their locally grown ingredients. However, this one could be classified as an "oaked saison" if the powers that be would have it. Not only is it aged for five months in Rioja casks, but it is fermented with a strain of yeast isolated from the bark of oak trees. This is unique territory, even for a saison.

What beer would be better to age than one that has been twice oaked? It has already been rounded and is good to go, but a further six months from when you shut the door and kick off your shoes wouldn't hurt. In fact, it would allow the wildness of the yeast to have its say. The bitterness would mellow, the woodsiness would come to the fore. It would be a fascinating few months ahead for the beer and you.

DOG EAT DOG 2.0

2007
BREWDOG
PUNK IPA

BEER & FOOD PAIRING

THE SCIENCE OF WHY IT WORKS

In times gone by, beer and food just went together. There was little thought as to the how or the why, only the when; food was provided and ales were consumed. Now, though, we live in the era of craft beer. The times have changed fundamentally. Science has stepped in and filled in the how and the why, leading us to an age of knowledge and an appreciation of which foods go with which beer styles.

More specifically, it is the sensory and flavor sciences that have done the heavy lifting. The impact of these branches of science has made the study of them a central component of the modern brewing industry. Their influence affects the decisions of brewers from the moment their beer is created to the minute before it is packaged – and even long after it has left the brewery.

Take yeast. Early brewers had no idea how it worked, believing their ingredients came together to become beer thanks to divine intervention. We have gone from that to a point where the study of yeast biology has determined its effects on the chemistry of aroma formation. In short, we now understand how different strains of yeast will deliver distinctive beer flavor profiles.

This is a quantum leap, and it is not alone. Sensory science exists to explain the perceived flavors arising from individual or grouped flavor compounds, and has resulted in all kinds of astonishing discoveries.

One such breakthrough was the knowledge that the active compound in chili peppers (capsaicin) triggers the same pain receptors on your tongue that bitter alcohol does. Each sends the same message to your brain that not all is well, and that you may have just ingested something harmful. This is anecdotal, if not definitive, proof that India Pale Ale makes spicy food seem hotter; they are both working the same pain receptors and your brain is getting the message in stereo. Of course, many people find this perception a pleasurable one. To each their own, we say, and if you enjoy the doubled-up "pain" sensation of spiciness, then more power (and antacid) to you.

A similar enhancement of flavors, and perhaps pleasurable to more people, is how the malty, sweet caramel flavors of a brown ale enhance the salty creaminess of most cheeses.

These are some examples of how contrasting

and harmonizing flavors found in different foods result in differing sensations and, therefore, enjoyment. Science is there to guide us in our quest for pleasure. So, let's use it to guide us right now.

The power of science lies behind all of the concepts of beer and food pairing, from the most basic to those that fly out of the park. Dig deep into their back story and you'll discover exactly why beer matches with food every bit as much as wine (if not more so), or anything else you can uncork, open, or pour when you plate up.

Let's start with intensity.
This notion is the simplest of all – light, delicate foods go best with lighter, refined beers. By that we don't just mean low-alcohol pale lagers, but a beer with subtlety and finesse in its flavor profile, unfettered by multiple ingredients or complex processes such as barrel-aging, which creates layered flavors. Most food pairings work because of complementary flavors, and they are so because they share similar aroma and/or flavor compounds.

SCIENCE IS THERE TO GUIDE US...

Take 4-Hydroxy-5-methyl-3(2H)-furanone. This flavor volatile is one among many that form when barley is kilned into malt. Each has a slightly different character, but that of 4-Hydroxy-5-methyl-3(2H)-furanone is evident when you come to know its layman's name – the "toffee furanone." So, when a stout that is full of sweet, roasted caramel flavors is drunk alongside a food sharing similar flavors (for example, chocolate cake), it complements the cake because they both have similar flavor compounds. The "toffee furanone" is the key to this ideal and complementary pairing.

Flavor compounds like this are at the heart of another, very different kind of result – the contrasting pairing. In a world where like goes with like, like can also go against its opposite. We touched on this briefly with the example of the heat of spicy foods being affected by the bitterness of hops, but we can indicate its power with that same piece of chocolate cake. Instead of cracking open a rich, dark stout to go with it, if you served up a Belgian Kriek, you'd be enhancing the bitterness of the chocolate but also the nuttiness of the cake as well. This is due to the volatile benzaldehyde (C_6H_5CHO,) which, amazingly, smells of almonds. The tartness of cherries will cut through the richness of the chocolate cake, cleansing the palate for the next mouthful, making it all the more... well, moreish.

And, at the end of the day, isn't that the reason we choose something to drink at mealtimes?

So, let's continue using science to our advantage and really harness the power of those tastebuds and aroma receptors. We are convinced that beer goes together with food better than anything else you can eat and drink at the same time, and we are about to prove it. Well, you are, but with our help.

In this chapter, we've laid out a series of multi-course beer-and-food-pairing dinners you can create at home. Invite your friends, family, and admirers – whether craft beer veterans or virgins. We know that pairing can be a thing of beauty, but the most effective way to demonstrate it is to build. Build layers. Create tiers of wonder. To be a true flavor geek, you need not one but many pairings. So, here they are.

Each menu that follows revolves around one of the basic principles of the art, laid out in triplicate: Starter, Main, and Dessert. To go a step further, we have selected a theme for each menu, based around one of the great cuisines of the world.

Have at it. Invite. Create. Open beers. Open minds. Food-pairing nirvana awaits.

BREWDOG

BREWDOG
JACK
HAMMER
RUTHLESS INDIA PALE ALE

BREWDOG

INTENSITY
MATCH OR CONTRAST THEME:
FAR EAST

When it comes to multi-course menus, jotting ideas down is the easy part. Cooking everything and making it work as a conveyor of Good Times is where the stress comes in. But don't sweat it – this is elevated beer-and-food matching, sure, but we aren't shooting for the Michelin stars. The best beer and food is enjoyed with friends, in as relaxed a manner as you can manage (and cooked with a beer at hand).

Let's begin with a slam dunk, a meal that celebrates great food, and even better beer, and illustrates the most fundamental pairing concept of them all: intensity.

Our first dinner involves choosing the correct degree of flavor to benefit both dish and glass, and it does so by increasing intensity in turn. Match a subtle flavor to a subtle flavor to begin, then move onward and upward. End with a big beer, a big dessert, and a greater appreciation of this cornerstone of beer-and-food concepts.

So, what area of world cuisine is best to demonstrate rising intensity in a meal? The one that marries delicate, floral dishes to gunpowder-strength combinations of chili, garlic, and ginger that you can taste for days, of course. The Far East has intensity locked and loaded; you'd expect nothing less from the birthplace of the fifth taste, umami.

INTENSITY OF THE FAR EAST

COURSE 1: LIGHT WITH LIGHT

Vegetable Tempura
paired with BrewDog Lost Lager

COURSE 2: MEDIUM WITH MEDIUM

Salmon and Miso
paired with Orval

COURSE 3: HARDCORE MEETS HARDCORE

Black Sesame Ice Cream
paired with Lervig Konrad's Stout

INTENSITY OVERLOAD

Once the meal is underway, this is your chance to see how powerful intensity can be. Try a piece of veggie tempura with a sip of Konrad's. Can you taste anything apart from coffee? In the battle of Light vs Hardcore there is only one winner...

VEGETABLE TEMPURA & BREWDOG LOST LAGER

COURSE 1: INTENSITY
LIGHT WITH LIGHT

The first course in this three-step hierarchy of intensity revolves around the principle of "less is more." When you think about beers that are light and delicate, you inevitably think of the lager family. And if you do the same with Asian cuisine, chances are your mind drifts toward Japan. Gently seasoned, thoughtfully prepared, and artfully served, Japanese food delivers classic, nuanced flavors that work perfectly with the classic, nuanced flavors of a Pilsner. Plus, fried food and lager is a no-brainer.

Before cooking, crack and pour your Lost Lager into a clean glass and take in the vibrant citrus and stonefruit notes from the dry-hopping regime of our 21st-century Pilsner. The German hops and Bavarian yeast have laid down foundations that are perfect for tempura, both as a pairing and as an ingredient, so use the lager as the liquid for the batter. Sparkling water also works, but beer gives a more perfect match and levels out the intensity of each. As a starter to your meal, this is the perfect beginning in every sense.

It is also seriously easy to prepare, which is another bonus. The first prep stage is to chop a selection of vegetables into bite-sized pieces. Onions, peppers, sweet potatoes, eggplants, zucchini, sugar-snap peas, and broccoli florets are all good. Whatever you can find and everyone likes, really. There's always one vegetable someone won't eat. Leave that on the shelf – or include it to demonstrate that everything tastes better battered.

SERVES 2

Ingredients:

vegetable oil, for deep-frying
2¼oz (60g) plain flour, sifted
½ teaspoon corn flour
2¾fl oz (80ml) BrewDog Lost Lager
About 7oz (200g) seasonal vegetables (we used eggplant, sweet potato, broccolini, asparagus, red pepper), cut into bite-sized pieces of equal size
salt and pepper
lemon wedges and a light salad, to serve

Heat enough vegetable oil for deep-frying in a large heavy saucepan over a medium-high heat to 350–375°F (180–190°C) or until a cube of bread browns in 30 seconds.

Combine the flour, corn flour, and a pinch each of salt and pepper in a large bowl, then pour in the lager and whisk with a fork to a thin batter. (Typically, tempura batter is made in a ratio of 1:1 flour to liquid, but the slightly larger quantity of lager here keeps it thin, rather than chunky and cloying. Use a fork to mix rather than blasting away with a mixer, as you don't want gluten from the flour to make things gloopy.)

Dunk the veggies in the batter until well coated.

Fry a few pieces of tempura at a time in the hot oil until golden, then remove with a slotted spoon to drain on a paper towel. Keep warm while you fry the remaining tempura.

Serve with lemon wedges, a light salad, and the rest of the Lost Lager.

PAIRING NOTES:

The crisp, cold Pilsner has the carbonation to clean the palate after every mouthful of batter-based goodness. Plus, the dry, bitter finish helps quell the richness from any oil attached to the tempura. The flavors will vary depending on what vegetable is up in your bowl next – but this will give a great viewpoint as to how individual elements affect the taste of a beer.

OTHER BEERS TO TRY:

- Bitburger Pilsner
- Fierce Beer Fierce Pilsner
- Oskar Blues Mama's Little Yella Pils

ORVAL

MISO SALMON & ORVAL

COURSE 2: INTENSITY
MEDIUM WITH MEDIUM

Move on to the main and the midpoint of our intensity pyramid, at which point we introduce the power of fermentation. All beer is fermented, but not all beer is fermented equally – and to get a better idea of what that means, Belgian styles are the ones to pick. Orval is a global classic and its earthy, herbal, funky edges are the product of mixed fermentation; yeast added by the brewers is augmented by the wild yeast brettanomyces. This isn't unique, but Orval is. To pair with it we are taking another product of fermentation – the mighty miso.

Orval is a thing of beauty – burnished gold, giant head, reducing to a mystery-clad lacing. Find your best glass and enjoy it before the work starts. Actually, the secret to this dish lies in the time given to the salmon to marinate – so maybe sort that out first and get it in the fridge before you head back for the Orval. Once the fish is taking on flavor, you can take some on yourself. This beer is a powerhouse when it comes to pairing, but one that equally works wonders by itself.

SERVES 2

Ingredients:

1½ teaspoons soy sauce
1½ teaspoons mirin
¾ teaspoon rice wine vinegar
1½ teaspoons water (or as needed)
1 tablespoon red miso paste
2 salmon fillets
6oz (180g) white rice
2 spring onions, finely chopped
1 tablespoon sesame seeds

In a bowl, mix the liquids and miso paste together into a fragrant, spoon-coating whole. If the liquids have the upper hand you can add less water so the marinade isn't too runny, but trust your eye and your taste buds to balance all the parts out into the greater good. Once combined, add the salmon fillets and let the sauce seep into the fish for at least 15 minutes.

Preheat the oven to 350°F/ 180°C and line a baking sheet with baking parchment.

Once the fish has taken on the flavor – and the color – of the miso marinade, place it on the prepared baking sheet and bake for around 15 minutes (the exact time will depend on the cut and size of your salmon, so keep checking; the fish will flake easily when cooked). The marinade that spilled onto the paper will dry and become crusty; the rest will keep the fish amazingly moist.

Meanwhile, cook some rice to go with it, using the failsafe 2:1 method. Tip the rice into a coffee mug and slide your finger down the inside until it makes contact. Keep your finger there and tip the rice into a saucepan, then head to the sink and fill the mug with water until it reaches your fingertip. Add to the pan, then repeat. Your finger measurement gives 2 parts water to 1 part rice. Cover the pan with a lid, bring to the boil over a high heat, then reduce the heat to low and cook with the lid ajar until all the water has gone. Fork through the rice to separate and leave the lid on for a bit to steam.

Serve the salmon on a bed of the rice and throw over the chopped spring onions and sesame seeds to garnish.

PAIRING NOTES:

The peppery, earthy elements of the beer work wonders with the spicy miso, while the sweetness from the malt bill merges into that from the mirin, not to mention the salmon. Orval is famously carbonated, which cuts through the fatty fish, and the fruity, funky parts of the beer counteract the sharpness of the rice wine vinegar and complement the umami of the miso. A winner.

OTHER BEERS TO TRY:

- Orval
- Orval
- ORVAL!!

LERVIG

BLACK SESAME ICE CREAM & LERVIG KONRAD'S STOUT

COURSE 3: INTENSITY
HARDCORE MEETS HARDCORE

Let's cut to the chase – this is one of the weightiest beers to ever come out of Norway. Weighing in at 10.4% ABV, Konrad's Stout pours blacker than the deepest reaches of any fjord. Lervig have been brewing it for over a decade, making it also one of the most consistently excellent imperial stouts in Europe. As such, it is a perfect way to round off our meal of increasing intensity. For this pairing, it shares the stage with sesame – black sesame seeds are nutrient-rich, powerfully flavored, and often found in Asian desserts. They give a one-two punch with the stout that takes humble ice cream to the next level.

Open the stout, take a deep sniff and a step back. The aromas once in the glass are like the strongest dark-coffee fruitcake you've ever encountered – only more intense, somehow. If you have chocolate cake lying around, Konrad's Stout makes a great pairing for it, but we are going in the other dessert direction and whipping up a batch of ice cream. Don't sweat about splashing out a three-figure sum on an ice-cream maker. All you need are a few ingredients, a food processor, a touch of elbow-grease, and time to spare for your freezer to do the work.

First, toast the black sesame seeds in a dry frying pan over a low-medium heat until you can smell them and hear them crackle and spit. Don't go too crazy or they'll burn, which is crucial to avoid but obviously hard to see. Blitz the warm seeds in a coffee grinder or food processor until they form a paste, then add the condensed milk and vanilla seeds or extract. Keep blitzing until everything is mixed together, then set aside.

Spoon the cream into a bowl and have at it with a whisk. Keep beating and whisking until soft peaks appear, then add a large spoonful into the sesame mixture and stir until mixed together. Add all the sesame mixture to the whipped cream and whisk again – you want everything the same color, albeit with tiny black specks of vanilla and sesame floating through.

Pour the mixture into a metal loaf tin and freeze for 6 hours.

SERVES 2 (WITH PORTIONS TO SPARE)

Ingredients:

3½oz (100g) black sesame seeds

14fl oz (400ml) can sweetened condensed milk

1 vanilla pod, split lengthways and seeds scraped (or 1 teaspoon vanilla extract)

14fl oz (400ml) heavy cream

PAIRING NOTES:

The key to this pairing is the roasted seeds, which mirror the roasted malts in the beer. The intense alcohol warmth is offset by the cooling ice cream, and the bitterness of the beer counters that famed condensed-milk sweetness. Plus, the coffee in the beer gives a roast 'n' toast element to the pairing. It works on so many levels.

OTHER BEERS TO TRY:

- Dieu Du Ciel! Péché Mortel
- Flying Dog Kujo
- North Brewing Co. Imperial Stout

THE BAMBERG ONION

One of the amazing things about the modern craft beer scene is that people are willing to travel for beer. Holidays, vacations, road trips, hell–even conferences are now booked in destinations because of the local beer. It wasn't that long ago that if you ended up somewhere with a great brewery, it was just a happy accident. Now, the suds call the shots.

Throughout this chapter we dial in to five of the world's greatest unsung dishes, and why you should make the effort to try them in their own backyard. All it takes to make it happen is a credit card and the click of a mouse. The first of these food odysseys revolves around a vegetable that isn't so much unsung as never even discussed:

The Onion.

But this isn't just any onion–this is the Bamberger Zwiebeln.

Seven thousand years of cultivation have led to this–the Franconian speciality that elevates the humble allium to near-godlike status. Served in gothic beerhalls in Bavarian backstreets, this is one for fans of slow food, hearty food, and comfort food. If you have ever had a stuffed vegetable before, this will leave you with an empty feeling about all of them. Eat one, and it won't just be the onion that is full to bursting.

Bamberg Zwiebeln are grapefruit-sized onions, hollowed out and stuffed with a mixture of pork, smoked pork, bread crumbs, eggs, herbs, and spices, then topped with smoked bacon. Over the top will be poured a gravy made from beef stock, the onion stewing liquor, and rauchbier. Alongside will be a mountain of mashed potatoes. The result is a lunch you won't skip away from. This is both-hands-on-the-table-to-get-up dining.

BEER PAIRING

Smoked Märzen. If you tell the story of your Bamberger Zwiebeln and someone wonders why you are drinking beer and eating a single onion, tell them a tale of the sweet, smoked beer and the spiced pork. Regale them with herbs, creamy mash, and the fragrant, caramelized onion merging with the malt-rich, sweet smoked beer. Watch the words falter in their mouth as they realize that you are serious and that this is worth booking a plane ticket for.

FLAVOR

MATCH OR CONTRAST THEME: MIDDLE EAST

Now that we're in the swing of things, the next time your beer-curious friends come a-calling, you can step things up with a meal revolving around flavor. Ideally, every single thing you cook revolves around flavor (or they won't come a-calling much), but we are talking flavor with a capital F here, and it illustrates another key fundamental of beer-and-food matching.

As with the previous menu, we are talking dinnertime science, not rocket science. The principle is very simple: if you have a flavor in a dish that matches up with a flavor in a beer, both will work well together and everyone will be happy.

Now, the problem – as you well know – is that neither "food" nor "beer" contains just a single flavor. They are all packed with all kinds of different things, giving your tastebuds a workout that's more than just a long boring treadmill into nothingness. But that's fine, we can pick out main themes and channel them in the same direction so that everything sings in harmony. And that's flavor.

Speaking of flavor, consider the food of the Middle East. Harissa. Pomegranate. Dates. Sumac. Lemon. Know any beers that share some of these? That's where the fun starts. This pairing dinner showcases a single note of a beer, does the same for a dish, and then raises the stakes with the same flavor derived from a process in both. All three work on one level and many levels at the same time.

MIDDLE-EASTERN FLAVOR

COURSE 1: BEER FLAVOR MATCHED TO A DISH

Flensburger Pilsener
paired with Falafel

COURSE 2: FOOD FLAVOR MATCHED TO A BEER

Sumac Kofte
paired with Brooklyn Sorachi Ace

COURSE 3: TECHNIQUE MATCHED TO A PROCESS

Harissa Lamb
paired with Aecht Schlenkerla Rauchbier Märzen

FLAVOR MISMATCH

Is this a concept? Well, yes and no. By all means, reel at the attempted pairing of smoky, rich Harissa Lamb and crisp, clean Flensburger; they shouldn't work. And yet they do, thanks to the power of Contrast. The bitterness of the beer cleans the palate and works wonders.

FLENSBURGER PILSENER & FALAFEL

COURSE 1: FLAVOR
BEER FLAVOR MATCHED TO A DISH

German Pilsners (or Pilseners) are more bitter and cleaner on the palate than their bready, sweeter-leaning Bohemian counterparts. But the single flavor we are teasing out today is that classic herbal note. Slightly grassy, mildly spicy, and, well, herbal, a lager that displays exactly this flavor comes from Schleswig-Holstein, Germany's northernmost state. Founded in 1888 and still in the hands of the same families, Flensburger Pilsener is the real deal. To illustrate its herbal edge, pair it with a dish that majors in fresh herbs: falafel.

This is a dinner you need to start the night before, but it's not a huge challenge. Soak the dried chickpeas in water overnight, 24 hours before you need to make the falafel. The big enemy here is consistency – using canned chickpeas is 1,000% easier but they add too much water, so the falafel fall apart.

Fast-forward from the previous night's soaking to tonight and pop the swing-top on the Flensburger. Take in the grassy, slightly bready aroma and then the bitter, herbal flavor. As the first "beer course" of a menu, it's the perfect style to start with.

SERVES 4

Ingredients:

14oz (400g) dried chickpeas, soaked overnight
1 small onion
3 garlic cloves
2 tablespoons chopped fresh parsley
2 tablespoons chopped fresh coriander
1 teaspoon ground cumin
½ teaspoon paprika
½ teaspoon ground coriander
½ teaspoon salt
a few twists of black pepper
plain flour, for thickening (optional)
vegetable oil, for deep-frying
salad, pickled turnips, plain yogurt, and pita bread, to serve

Drain the soaked chickpeas and dry them thoroughly on paper towel to remove as much moisture as possible. Transfer them to a food processor along with the onion and garlic and blitz to combine, then add all of the herbs and spices before giving it a second blast. Check to see how sticky the mixture is – you want to be able to form balls that will hold together, so the mixture can't be too wet. Add a little flour or extra water to adjust the consistency if needed. Leave the mixture to rest in the bowl for around 20 minutes.

Meanwhile, heat enough vegetable oil for deep-frying in a large heavy saucepan over a medium-high heat to 350–375°F (180–190°C) or until a cube of bread browns in 30 seconds.

Form the mixture into 30 golf-ball-sized spheres. Pick a sacrificial falafel and fry it in the hot oil to see how it does. If it browns, cooks through, and holds together, you're on. Fry the falafel in batches until golden brown all over. Remove with a slotted spoon to drain dry on paper towel.

Serve with pita bread, yogurt, salad, and the Pilsener.

PAIRING NOTES:

Flensburger is notable for its clean, herbal bitterness, and that mirrors the parsley and coriander perfectly. The Pilsener also works to cut the doughiness you can get if the falafel mix isn't cooked through enough, and does exactly the same for the pitta. It's the perfect palate reset for the next mouthful.

OTHER BEERS TO TRY:

- Firestone Walker Pivo
- Jever Pilsener
- Five Points Pils

SUMAC KOFTE & BROOKLYN SORACHI ACE

COURSE 2: FLAVOR
FOOD FLAVOR MATCHED TO A BEER

Of all the styles, saisons are one of the most versatile when it comes to food pairing. Their flavors can lie anywhere on the spectrum, and they are highly carbonated, dry, and bitter. This means they can stand up to pretty much anything, plus you can knock one back from a can or enjoy from the most precious of wine glasses. Saisons are, in short, it. So let's dive in with a unique example from Brooklyn that is reflected, near perfectly, in a tangy, lemony spice from the Middle East: sumac.

Open up Brooklyn's finest and enjoy this classic saison, enlivened to new heights thanks to the unique Japanese Sorachi Ace hops. It unleashes flavors as unusual – and complementary – as coconut, dill, and lemon zest, and creates a fantastic base layer for all the other flavors that are about to emerge in the dish you are preparing.

To create the kofte, simply put all the ingredients into a bowl and get mixing. That's pretty much it. If you are fixing these to go with the falafel in Course 1, they can even be cooked in batches alongside each other, fried in the same oil (although if you have vegetarians arriving, cook the falafel first).

SERVES 4

Ingredients:

1lb 2oz (500g) minced lamb
2 small shallots, finely sliced
2 garlic cloves, crushed
1 tablespoon sumac
1 teaspoon cumin
1 teaspoon black pepper
1 teaspoon salt
½ bunch of flat leaf parsley, finely chopped
1 egg
2 tablespoons fresh bread crumbs, or more as needed
vegetable oil, for frying
pita bread or salad, to serve

Prepping is easy: put the minced meat into a large mixing bowl, throw in the shallots and garlic, and mix together with your hands. Add the dry spices and parsley and continue to mix.

Add the egg and bread crumbs as required – they are there to bind the meat together so that it forms kofte that won't fall apart later. Too much egg and the mixture will become mulchy and super slippery, so a handful of extra bread crumbs will be required to balance it out. Trust your eye, and your hands, to work out what's needed as you knead the fragrant minced meat together.

Shape into 16 equal-sized sausage-shaped pieces and add to the hot oil. Brown on each side and then cook for 5 minutes on each side, until cooked through.

Serve in a pita or with a salad (pomegranate seeds are a great addition here) and enjoy with the saison.

PAIRING NOTES:

We'll start with the food this time – the sumac brings an airy, floral, lemony quality and the cumin a deeply roasty spice element. Lemon, citrus, and spice are all qualities possessed by the beer as well – the bright herbal edge of Sorachi blends perfectly with that of the kofte, and the spice from the Belgian yeast marries with the final flavors of every mouthful. With carbonation to keeps things going, this is matching at its best.

OTHER BEERS TO TRY:

- Saison Dupont
- Wild Beer Co Epic Saison
- anything by Hill Farmstead (if you can find it)

HARISSA LAMB & AECHT SCHLENKERLA RAUCHBIER MÄRZEN

COURSE 3: FLAVOR
TECHNIQUE MATCHED TO A PROCESS

Smoking isn't cool. But it can be. To illustrate how, we round off our flavor-high pairing dinner with the power of smoke, teased between food and beer. More specifically, we are going to play off two different types of wood: beech and oak. The latter is burned to flavor peppers, creating sweet but smoky pimentón (Spanish paprika). The former is burned to dry barley into rauchmalt for the classic smoked beers of Bavaria. There's no smoke without fire, and this pairing proves that twice.

Let's get smoked. Open up the Märzen and breathe it in. The rich, deep, bacony smoke aroma that arrives is the most distinctive bouquet beer has. Centuries of tradition in drying and smoking malt over beechwood has given these Bavarian exports their uniqueness. The flavors are, if anything, even more full-on. If you don't have the time for the lamb dish, then these beers also go brilliantly with hot dogs. We often say the beauty of beer and food pairing is the versatility beer has – this recipe proves it.

Preheat the oven to 340°F/ 170°C.

Coat the lamb cubes in the harissa and pimentón. Both of those pastes can be found in varying degrees of strength, but typically harissa is the more fiery due to the chilis. Using the sweet version of paprika balances the harissa out and brings a greater depth of flavor to get into and around the lamb.

Put the lamb into an oven dish, then add the tomatoes, onion, and tomato purée, plus a little water if it looks like it needs it. Cover with a lid or foil and bake in the oven for 1 hour. Remove and check how the sauce looks. If it looks too dry, add a little water to bring it back if necessary.

Stir in the chopped cilantro and serve immediately, with flatbread or rice (or both).

SERVES 4

Ingredients:

1lb 12oz (800g) lamb, cubed
1 tablespoon harissa
1 tablespoon sweet pimentón (or sweet smoked paprika)
14oz (400g) can plum tomatoes
1 red onion, diced
1 tablespoon tomato purée
small bunch of cilantro, chopped
flatbread and/or cooked rice, to serve

PAIRING NOTES:

So, the process of smoking malt and the technique of roasting peppers over a wood fire pay off when combined. You might think the result would be like breathing in campfire smoke, but there's a harmony rather than a clash because the smoke is both sweet and roasty. Just as with the balance of harissa and paprika in the sauce, these cousins born of fire can bind together and become greater when paired. Beech and oak give two distinct profiles to the beer and the spice, and they work beautifully when united.

OTHER BEERS TO TRY:

- Alaskan Brewing Smoked Porter (smoke source: alder)
- Aecht Schlenkerla Eiche Doppelbock (smoke source: oak)
- Stone Smoked Porter (smoke source: peat)

THE LOUISIANA PO'BOY

The Deep South almost has too many culinary specialties to countEven small regions are likely to have more than one, this varies depending on whether you are surrounded by farmland, rolling hills, or marshy delta. For a case in point, head to the mouth of the greatest river in the nation and the city of New Orleans. In the land of gumbo, boudin, jambalaya, and boulettes, you are looking for a sandwich. But calling this a sandwich is like calling the Bamberger Zwiebeln an onion (see page 130). This is po' boy country.

As befits a part of the world where influence came and went with the people bringing or losing it, this sandwich draws inspiration from the cuisines of several countries. A bit British, quite a lot more French, but with a uniquely American twist, the po' boy has grown from a backyard staple to a regional delicacy that can be filled with anything from duck to rabbit. But the original? Roast beef.

Long French bread, stuffed with hot roast beef and with gravy ladled over the top, was the way po' boys were, long before the seafood surge that today has given the Big Easy legion po' boy counters with shrimp, soft-shell crab, or fried oyster versions. These are all good – if you get a poor po' in Louisiana then you are truly unlucky – but when seeking out the real deal, a roast beef and gravy French bread sandwich is the Delta King.

These are subs that are less "meal deal" than full-scale meal. Doorstop wedges of crusty bread that arrive looking like a treasure chest unable to be fully closed, spilling its contents in front of you. They are loaded, stacked, and piled high. The beef is simmered for hours in stock until it falls apart with a hard stare, so the saying goes. Rich, moist, and impossible to leave unfinished, po' boys are the real deal.

BEER PAIRING

American Amber. American coppery-brown, malt-led beers are ideal bedfellows for anything that revolves around roast beef. The hit of sweet caramel and toasty bread crust from the malt mirrors everything you are about to consume, and the elevated bitterness from the American style gives you the perfect way to clean your palate before you clean your plate. Amber ales, Red IPAs, American brown ales – any of these work, and given the size of the average po' boy, you may need all of them.

HOPS

MATCH OR CONTRAST THEME: BEST OF BRITISH

Stop us if you've heard this before, but hops are amazing. The preservative of old has matured into craft beer's signature ingredient, and fortunes have been made as a result. That's because the list of characteristics conveyed by hops is limitless when they are deployed in combination – brewers now can tease out flavors and aromas you'd never expect to find in beer. And it's why we are taking it to the wine guys at mealtimes. Leave that corkscrew in the drawer.*

Let's say you want to host a pairing dinner to bring a tear to the eye of the most hop-phobic friend you've got. How do you convert the lupulin deniers to the power and majesty of the cone? Easy. You pick out a single facet of the versatility of the hop and then change it up three times in a row.

That facet: bitterness.

From cheek-scraping India Pale Ales to pints of the pub-friendly style named after the taste itself, bitter is one of the signature moves of the hop. It's also true that the sensation of bitterness is a great way to showcase how contrasting elements of a dish and a beer can bring out the best in both. Let's give it a try and turn to the land that made contrast a way of life (and not just in terms of the weather), the UK.

Lace up that Union Jack apron and show how bitterness lifts these home-grown British recipes to the next level.

*Unless your beer has a cork. We're not animals.

BEST OF BRITISH

COURSE 1: BITTERNESS CUTS FATTINESS

Grilled Mackerel
with Sierra Nevada Pale Ale

COURSE 2: BITTERNESS EMPHASIZES HEAT

Chicken Madras
with BrewDog Punk IPA

COURSE 3: BITTERNESS REDUCES SWEETNESS

Carrot Cake
with Almanac Loud!

BITTER ENEMIES

Bitterness, like everything else, is relative. The International Bitterness Units (IBU) scale rates beers from 1 to (the theoretical maximum) 100. Try a beer with low IBU and see how it tackles the fattiness of mackerel or the sweetness of carrot-cake icing. Not as well, most likely.

GRILLED MACKEREL & SIERRA NEVADA PALE ALE

COURSE 1: HOPS
BITTERNESS CUTS FATTINESS

Let's get contrasting! It's not often you get to create a dinner where everything is designed to cancel out something else (unless you are talking about the perfect guest seating plan), but that's what we're doing here. Actually, it's more countering than cancellation – we are still demonstrating a net gain. So, every time you eat something and think, "Oof, that's a bit...," there's a beer that can come to the rescue. Take fattiness. Rich, oily fish is much-needed for the body, but eat too much and you sit back, slightly queasy and full. Pale Ale is your savior, my fish-loving friend. Pale Ale.

Open the world-famous bottle that has launched the careers of thousands of brewers (and at least one writer) and be thankful that you live in a time when SNPA is also with us. When it's all said and done, Sierra Nevada Pale Ale, devised in 1980 and still more than relevant today, will be remembered as one of the most important beers in history. Get a few in and enjoy at least a couple before the mackerel hits the grill. It's the perfect first, warm-up beer and rapier to cut through the forthcoming fatty fish.

Mackerel are silvery, omega-3-rich, super-sustainable and are available locally in the northern hemisphere in summer and imported just about year round. Scaled, gutted, and pin-boned (either by you or by your fishmonger) they couldn't be easier to cook.

Preheat the grill to high.

Simply brush the mackerel fillets with oil and season with salt and pepper, then pop the fillets onto the grill pan and grill for 2 minutes on each side or until cooked through. That's it. With this fish, there's really very little else you need, other than a small drizzle of lemon juice before serving. For best results, char the skin side so that it's crispy and delicious.

Serve with a potato salad, mixed greens, and a fridge-cold Sierra Nevada Pale Ale. If you make only one thing from this book, make this.

SERVES 2

Ingredients:

2 mackerel fillets, scaled, gutted, and pin-boned
olive oil, for brushing
½ lemon, for squeezing
salt and pepper
That's it!

To serve:

potato salad
mixed salad leaves

PAIRING NOTES:

We've talked a little already about the concept of bitter beer cutting through fattiness – and this is the pairing to prove it. A couple of bites of fish and the mouth is coated with a comforting, oily fattiness (a sensation that can stay with you for hours if you eat a lot of fish in one sitting). Take a pull on the pale ale and – whack – it's gone. Your mouth feels refreshed, your palate cleansed, senses primed. This is bitterness cutting fattiness in action. Maybe have another bite though, to be sure…

OTHER BEERS TO TRY:

- Pressure Drop Pale Fire
- Cloudwater Pale Ale
- Upslope Pale Ale

PALE ALE
PALE ALE

CHICKEN MADRAS & BREWDOG PUNK IPA

COURSE 2: HOPS

BITTERNESS EMPHASIZES HEAT

Next up on this classic British menu – curry. The cuisine that contributes over five billion pounds to the UK economy is its real national dish and a useful tool to demonstrate one of the cornerstones of pairing (it's also delicious, let's not forget about that). Curry and IPA have been placed together ever since beer and food became a thing – and maybe not with the best results. As we are about to see, they do work together, but one compounds the other. Hop bitterness makes your curry hotter. Let's go.

Before you get started, open the beer that started it all for us. Punk IPA was inspired by the hop-forward beers of the US that our co-founders James and Martin fell in love with but couldn't find, prime among them the aforementioned Sierra Nevada Pale Ale. With the bitterness of the finest American hops, Punk IPA is spiky and tropical, with an explostion of grapefruit, pineapple, and lychee flavors. It also carries the bitterness to amplify our upcoming curry.

As with many popular British curries, our Madras might not be recognized in India. Not even the name is used there anymore, as the southern Indian city that gave its name to the chili-rich sauce was renamed Chennai in 1996. The ground spices for this dish can either be store-bought or home-ground (the latter will give you a greater depth of flavor), but be sure to have everything ready to go.

SERVES 2

Ingredients:

1 tablespoon vegetable oil
1 small onion, finely sliced
1 garlic clove, sliced
2 red chilis, deseeded and sliced
2 chicken breasts, chopped into bite-sized pieces
1 teaspoon garam masala
1 teaspoon ground coriander
1 teaspoon ground turmeric
1 teaspoon ground cumin
1 teaspoon ground ginger
14fl oz (400ml) can coconut milk (optional)
1½ teaspoons tomato purée
1 large tomato, chopped
3 tablespoons (50 ml) chicken stock

To serve

cooked basmati rice or naan

Heat the vegetable oil in a large saucepan, add the onion and fry until browned, then add the garlic and chili and cook for about 30 seconds until fragrant. Add the dried spices and cook for about 15 seconds to create a dark, heady mix of heat and flavor, then add the chicken, stirring to coat it in the spices. Cook, stirring, for about 5 minutes until the chicken is cooked through.

At this point, you can add coconut milk if you want to mellow things a little, but we just add the tomato purée, a chopped tomato, and some chicken stock. Simmer the curry until the liquid has reduced to your preferred consistency and the curry is ready to go.

Serve with some basmati rice (cooked using the method on page 127). Naan is great here too.

PAIRING NOTES:

This combo is one for fans of heat and spice. The bitterness of the Punk IPA brings the heat from the curry into sharper focus, and in turn raises the perception of the beer's bitterness. Each one raises and emphasizes the other. After a while, the tropical notes from the beer do bring the pairing back down to earth a little – but if you like things hot, try a curry and super-bitter IPA and see what they can do for each other (and for you).

OTHER BEERS TO TRY:

- Buxton Axe Edge
- Bell's Two Hearted Ale
- any IPA from the Kernel Brewery

CARROT CAKE & ALMANAC LOUD!

COURSE 3: HOPS
BITTERNESS REDUCES SWEETNESS

OK, so this is something you wouldn't expect. Carrot cake and beer? But take our word for it – if you want a single depiction of the influence beer can have on food (and vice versa) then find a piece of carrot cake and open a Double IPA. The high bitterness level in the beer cuts through the sweetness of the cake and its icing. It's like rock paper scissors: bitterness cuts sweetness. You may never have thought of beer and cake before, but it's actually a fascinating thing to try. And why buy carrot cake when it is so easy to make?

Snap open a can of Almanac's finest and pour into a clean glass. Take in the sticky pineapple and mango aroma and flavor, and particularly note the oily honeyed finish from the high alcohol and higher levels of hops therein. This is the sensation that will be going up against the cake you are about to make.

MAKES 1 LARGE CAKE (LEFTOVERS!)

Ingredients:

12oz (350g) plain flour
2 teaspoons baking soda
1 teaspoon salt
1 tablespoon ground cinnamon
10fl oz (300ml) vegetable oil, plus extra for greasing
5 large eggs
7oz (200g) superfine sugar
7oz (200g) light brown sugar
3½oz (100g) walnuts, chopped
1lb 2oz (500g) carrots, grated
zest and juice of ½ lemon

For the icing:

5½oz (150g) unsalted butter
7oz (200g) cream cheese
10½–14oz (300–400g) powdered sugar
2 teaspoons vanilla extract
zest and juice of ½ lemon

Preheat the oven to 350°F/180°C and grease two 9 inch (23cm) cake tins with a little oil. Have another quick drink of the beer before reaching for that wooden spoon.

Cakes are best made in 2 separate bowls of dry and wet ingredients before combining, so whisk together the flour, baking soda, salt, and cinnamon in one large bowl, then do the same for the oil, eggs, and sugars (sugar isn't wet but just go with it) in another bowl. Mix the wet ingredients into the dry, stirring to combine, then add the nuts, carrots, and lemon zest and juice. Stir well so that everything is evenly mixed.

Pour evenly into the 2 prepared tins and bake for 30 minutes, or until golden brown and a toothpick comes out clean.

Turn out to cool on a wire rack.

While the cakes are doing just that, make the icing. Beat together the butter and cream cheese and, when it looks nicely mixed, stir in the icing sugar, vanilla, and lemon zest and juice until well combined.

Spread the icing over, inside, and around the 2 halves of the cake, before joining together. Make it as neat or as messed-up as you like – it will still taste the same. There should be no judgments when it comes to baking.

PAIRING NOTES:

Reach for your DIPA and take a drink after your first couple of mouthfuls of cake. Snap. The sweet sugariness in your mouth should vanish before you put the glass down. It really reduces before your very tastebuds. The Mosaic and Simcoe hop oils that give bitterness to the beer moderate the sweetness of the cake and icing. Best take another mouthful of each in turn just to be sure, though.

OTHER BEERS TO TRY:

- BrewDog Jack Hammer
- Against the Grain Citra Ass Down
- Dogfish Head 90 Minute Imperial IPA

LOUD! HAZY DIPA
Almanac
BEER CO.

THE MORETON BAY BUG

Australia's eastern seaboard is lapped, splashed, and pounded by waters that contain a myriad of different creatures, only a few of which are deadly. Far more are extremely tasty to eat and, as befits an island continent, there are hundreds of local specialities that are hooked, netted, or dredged up from the teeming seas that you won't see in many other places.

One such local example are the flat, shovel-nosed lobsters known locally as Moreton Bay or Balmain Bugs. These two different species live similar lives buried in the sandy mud before emerging at night to roam around looking for food. Since they aren't rock-dwelling lobsters, they look as if they have been smooshed a little, flatter and spread out, without the formidable claws clacking away in their defence.

A common sight in seafood restaurants on the coast of New South Wales and Queensland, bugs are often grilled, steamed, or poached. They are also amazing when barbecued (although, without resorting to stereotypes, "throw another bug on the barbie" might not catch on). No matter how they are cooked, they should be eaten within view of the sea – this is imperative. Bugs are seaside cuisine.

Bugs are also far easier to tackle than their lobster cousins. The beauty of the bug is that without claws or other tough-going appendages, all the meat is in the tail. The person doing the prep will have split them, so all you need to do is pull out the wedge of sweet tailmeat and enjoy. They have a stronger, but still sweet, seafood flavor that borders on the fishy. They are both incredible and incredibly addictive.

BEER PAIRING

Pale Ale. The depth of flavor these flat, slipper-shaped lobsters have means you can pair them with a beer that brings more oomph than members of the lager family. The elevated citrus notes of a classic Australian pale ale makes an ideal match – the carbonation removing any buttery fat residue and the sweetness of the meat matching the pale malt at every turn.

MALT

MATCH OR CONTRAST THEME: US OF A PARTY

Albert Einstein never wrote about beer, that we know of, but if he had his theory of beery relativity would hold that for every article about hops, an article about malt should follow to balance it out. Craft beer has changed that, of course, as the modern worship of hops has put the universe out of whack – but let's be true to Albert's teachings and work up a pairing dinner that demonstrates just how powerful the backbone of beer can be.

The humble grain has a lot going for it, particularly when picking up the pencil to create a pairing dinner. In fact, if you think of the ways in which malt affects a beer, and can then affect food, we'd argue that malt is even more pivotal than hops. If you gasped in disbelief at that, then sit down to this meal.

Before we start rattling pans, we should – of course – state the obvious that "malt" is a catch-all term for a number of different cereal crops, processed to aid brewers. As a result, instead of a single secret, it really has an entire suite of game-changing attributes. Let's pick out three big ways in which you can use malt-derived flavor to your advantage when it's time to eat.

We can take a trio of classic American dishes and raise each one to the roof thanks to the power of malt.

US OF A PARTY

COURSE 1: ROASTINESS TAMES TARTNESS

Cobb Salad with Blue Cheese Vinaigrette
paired with La Trappe Trappist Dubbel

COURSE 2: SWEETNESS BALANCES SPICE

Five-Alarm Chili
paired with Anchor Old Foghorn Barleywine

COURSE 3: SMOOTHNESS BLANKETS BITTER

Key Lime Pie
paired with Ayinger Ur-Weisse

THE MEDLEY OF MALT

Our contrasting malt meal could just as easily be a matching malt meal – it's no coincidence that several malt varieties sound like menu items: Chocolate malt. Caramel malt. Biscuit malt. Wheat. Oats. Compiling a pairing dinner with beers and food that share these flavors is another fundamental of the art.

COBB SALAD WITH BLUE CHEESE VINAIGRETTE & LA TRAPPE TRAPPIST DUBBEL

COURSE 1: MALT
ROASTINESS TAMES TARTNESS

What's tart about a salad? Well, like all salads, the answer is: it's up to you. The best salads are thrown together and have all kinds of flavors, textures, and layers. So, when pairing with beer they can be a tricky prospect. But we did say malt is more pivotal than hops in food matching, so we'd better walk the walk. When you need the big guns, you can rely on the Order of Cistercians of the Strict Observance. Trappist beers are all that and more, particularly for a salad containing blue cheese, watercress, and spring onions, laced with vinaigrette. Sharp, biting salads are tamed by Trappist wonder.

Fruit, caramel, dates, vanilla, chocolate, honey. Not the salad; these are aromas that can be found emanating from a freshly poured glass of La Trappe Dubbel. This is one of the classic Trappist ales of the world, and is hugely versatile when it comes to food pairing. The only difficulty is that if you pour it first (as we advise with all with all pairings), it's likely to end up disppearing before it comes time to plate up and put it to the test. Have a taste and stay your hand – there's chopping to be done.

SERVES 4 AS A STARTER

Ingredients:

150g (5½oz) pork lardons
3 skin-on chicken breasts
4 large eggs, hard-boiled
3 large vine-ripened tomatoes
1 romaine lettuce
1 bunch of watercress
6 spring onions
1 avocado
100g (3½oz) blue cheese (Roquefort is best)

For the vinaigrette:

4 tablespoons red wine or sherry vinegar
2 teaspoons Dijon mustard
150ml (5fl oz) olive oil, or as needed
salt and pepper, to taste

First, fry up the lardons in a dry frying pan until they are crispy, then add the whole chicken breasts to the pan and cook for about 5–6 minutes on each side until golden brown and cooked through – the bacon flavor will give them even more than their own. Remove from the heat and let cool, then slice the chicken breasts.

Meanwhile, peel and quarter the eggs, chop the tomatoes into wedges, and roughly chop the lettuce and watercress. Slice the spring onions, halve the avocado, and then slice that too. Finally, break the cheese into small nuggets (the knife has done enough).

Arrange all the salad ingredients on a large plate. Cobb salad is named after the Los Angeles restaurateur who invented it, apparently as a "what's left in the kitchen?" evening meal – and the traditional way to present it is to have everything lined up in sections on top of a bed of the lettuce, rather than all jumbled together. But hey, it's up to you. Add to the plate as you see fit.

Make up the dressing: whisk together the vinegar and mustard in a bowl, then whisk in enough of the oil to balance out the sharp hotness. Once happy, season to taste, pour over the salad, and reach for what's left of your Dubbel.

PAIRING NOTES:

This first malt reaction is an exercise in subjugation. A strong, dark beer with a salad seems odd. But this salad has so much bitter, sharp, tartness from the ingredients that this is precisely what you need. A lighter beer would taste like water. The sweet, full, malt-led roastiness tames the tartness and complements it perfectly. It allows the piquancy to come through, but only just enough. It really works.

OTHER BEERS TO TRY:

- Westmalle Trappist Dubbel
- Allagash Dubbel Ale
- Lost Abbey Lost and Found

D
Trappe
Dubbel

FIVE-ALARM CHILI & ANCHOR OLD FOGHORN BARLEYWINE

COURSE 2: MALT

SWEETNESS BALANCES SPICE

Let's go back a little, to where we demonstrated that the bitterness of hops makes hot food seem hotter. What's the alternative? Well, either you like that kind of thing, or you reach for a barley wine. Any beer that has a malt-derived sweetness, packs caramel malt, or has a high ABV and low IBU can come to the rescue. Barley wines (or barleywines; either is good) are deep, strong vectors of flavor that bring back dishes that would otherwise leave you at the brink. Hot, spicy, pungent food is balanced by multi-malted beers. Don't take our word for it – tame this lightning-hot chili with the best barley wine in the world.

The hotter the fire, the more alarms it sets off – and this chili is the big one. Thankfully, San Francisco's finest are here to save the day, so before you lay a path leading to those ringing bells, open a bottle of Old Foghorn Barleywine. First brewed in 1975, this ageless beer peaks from two sources – multiple additions of Cascade hops and the power of caramel malt. The stewing process that leads barley grains to become sweet and honeyed is the secret to balancing any amount of spice. Even five-alarm levels.

Note: Feel free to reduce the number of alarms if you need to, working backward. Lose the chili flakes first, use kidney beans in water rather than chili sauce, and so on. Keep going until you have the heat under control.

SERVES 4

Ingredients:

1 tablespoon vegetable oil
1 onion
2 garlic cloves, chopped
4 streaky bacon rashers
1lb 2oz (500g) minced beef
1 tablespoon chili powder (first alarm)
2 habañero chili peppers, chopped (second alarm)
1 tablespoon ground cumin
1½ teaspoons cayenne pepper (third alarm)
2 x 14oz (400g) cans chopped tomatoes
14oz (400g) can kidney beans in chili sauce (fourth alarm)
1 teaspoon malt vinegar
1 tablespoon tomato purée
1½ teaspoons dried chili flakes (fifth alarm)
9fl oz (250ml) water

To serve:

cooked white rice
tortilla chips
sour cream

Heat a little oil in a large saucepan, add the onion and fry until translucent, then add the garlic and bacon and fry until cooked through. Pour out into a bowl and set aside. Add the minced beef to the same pan and cooked until browned all over, then add the chili powder, habañeros, cumin and cayenne pepper and stir until combined. Add the bacon and onion mixture back into the pan, then stir in the tomatoes, kidney beans, vinegar, and tomato purée. That's pretty much it – let the chili cook down and increase in flavor and intensity, adding the water little by little to keep it from drying out too much and to increase the amount of sauce you'll get.

When it looks nearly done, hit that fifth alarm by sprinkling in the dried chili flakes. Once the chili is reduced to your liking, serve with rice, tortilla chips and plenty of sour cream. And the Barleywine – you will need that within reach.

PAIRING NOTES:

As you undoubtedly know, the heat of something like a chili keeps on rising and peaks long after you've put your fork down – that moment is when the beer will help you out. The sweetness of the caramel malt will be amplified by the heat and will help balance it all out. The spice will become roastier. The sweet malt and browned meat will sing in unison. The alarm will fade into deliciousness.

OTHER BEERS TO TRY:

- Swannay Old Norway
- AleSmith Old Numbskull
- BrewDog Clown King

KEY LIME PIE & AYINGER UR-WEISSE

COURSE 3: MALT SMOOTHNESS BLANKETS BITTER

Everyone needs a blanket every now and again. Here, though, this classic dessert is not the blanket, but the reason why one is required. Bitterness doesn't only come from hops. Way above them on the cheek-sucking scale is the humble lime. The savior of your G&T also makes rich, creamy, mom-and-pop-diner pies bitter and sharp – so, to counter it in this final act of our malt-led contrast meal, reach for a Dunkelweizen. This German style is sweet, rich, and smoooooth. *This* is your blanket.

Open your Dunkelweizen. Maybe whisper the name of the style out loud a few times as you do. Essentially it means "dark wheat beer" and when you pour, you'll see why. A mysterious dark amber beer appears in your glass, sporting a massive foam head. The aroma is sweet and earthy, with some spice, toasted nuts, and caramel. All perfect foils for a bitter, sharp lime dessert..., which we should get around to making.

SERVES 4

Ingredients:

9oz (250g) unsweetened Graham crackers
1¾oz (50g) flaked almonds
2 tablespoons superfine sugar
3½oz (100g) unsalted butter, melted
3 large egg yolks
14fl oz (400ml) can condensed milk
5fl oz (150ml) heavy cream
zest and juice of 4 limes, plus extra lime zest to decorate

Preheat the oven to 350°F/180°C.

Put the graham crackers and almonds in a sealable bag and have at it with a rolling pin, meat tenderizer, or even the empty Ayinger bottle. Crush them up good. Add the sugar, then pour the bag's contents into a bowl and add the melted butter. Stir to combine into a crumb and use this to line a 9 inch (23cm) pie dish or cake tin. Bake for 10 minutes until the crust is even more golden, then remove from the oven and let cool.

Meanwhile, put the egg yolks into a mixing bowl and use a hand-held electric mixer to beat them together for a minute, before adding the condensed milk. Continue beating for a while, then add the cream, then the lime zest and lime juice. Keep mixing for a few more minutes until everything is nicely soft and creamy. Once the crust has cooled, pour the filling into it, and bake for 15 minutes. You want it to be set, so give it a shake after the time is up and check it's all good. If not, put it back in the oven for a short while.

Chill the pie in the fridge overnight and decorate with more lime zest, if you have it.

PAIRING NOTES:

Key Lime Pie has two classic flavors: the sharp, bitter lime and the soft, creamy sweetness of the filling. That's what our beer is here to react to, as the wheat-derived sweetness blankets the bitter part of the dessert, while the darker malts used in the brew complement the graham cracker crust perfectly. The dark weizen has enough bitterness to carry the sweet filling, yet counter the moments of lime zest encountered.

OTHER BEERS TO TRY:

- Weihenstephaner Hefeweissbier Dunkel
- Schneider Weisse Original
- Maisel's Weisse Dunkel

Ayinger
URWEISSE

STORNOWAY BLACK PUDDING

At Scotland's northwesterly extreme lie the Outer Hebrides. On the map, they look like an arm raised in protection from the elements, as if to deflect the many storms and squalls that roll in off the Atlantic Ocean. The largest settlement here was founded by the Vikings, and bears a name that sounds almost descriptive of their attempt to keep bad weather at bay.

Stornoway is a hard place to reach, given that it is perched on the edge of an island, but its exports have reached further afield. One of the most famed of these is Stornoway black pudding. Typically eaten at the start of the day rather than after a meal, this kind of pudding is a dark, hugely flavorful blood sausage, served sliced and fried and tasting utterly unique.

The real deal, made only on the Isle of Lewis, contains beef suet, oatmeal, onion, blood, salt, and pepper. The blood is either sheep's, cow's, or pig's, and this is what sets Stornoway's local version apart from others (which are likely to be pig's blood). This underpins its history: life was hard in the Hebrides, and when any animal was slaughtered for meat, they had to use every part of it, even the blood.

These days, the plastic-wrapped pudding appears in butchers and delis around the UK and beyond–if you think Scotland is renowned solely for haggis, think again. Black pudding is savory, iron-rich, spicy, and full of umami depth. It is far more than an addition to a Full Scottish Breakfast and works in all kinds of dishes – use it in place of chorizo for a world of flavor, straight from the periphery of Scotland.

BEER PAIRING

Saison. The obvious approach here would be a stout: black pudding, black beer. The oatmeal in the pudding works really well with a similarly malted stout, but for a fascinating act of contrast, try a saison. The pepperiness and spice in the pudding mirror that in the Belgian yeast, but the sweet citrus of the beer counters the massively rich palate overload from Stornoway's finest. An unexpected majesty.

SAUCE

MATCH OR CONTRAST THEME: FROM THE UK TO THE WORLD

We often talk about the wonders of barrel-aging, and how putting a beer into oak can fundamentally change the character, flavor, and strength of the final product. When heading into the kitchen, something that does exactly the same for a dish (which is often overlooked when it comes to beer-and-food pairing) is the sauce. Whether from a jar or a classically trained *saucier*, sauces take your dinner in a different direction.

That's their intention, of course. From whimsical foam to ladles of thick, rich gravy, adding a sauce brings a contrasting edge of flavor to the main meal it envelops. And that can cause problems for even the highest-rated beer-and-food pairer. When drawing up a menu, always consider the sauce, as that final arrival can knock a pairing off the plate or out of the park.

Let's demonstrate that principle with a trio of sauces commonly found but hard to tackle. Feel free to create and deliver these as a meal, although – as you can see – it would be a menu of mains. A better option would be a tasting menu of all three. For a fuller course affair, select one from this section, then pick from our final menu (Cheeseathlon, page 163) for the second course (no peeking).

To demonstrate the power that sauces convey, we are sticking with a single country when it comes to the beers. Feel free to substitute if needed, but the styles are key to taming these three tricky, sticky sauces.

FROM THE UK TO THE WORLD

SAUCE 1: BARBECUE

Spicy Bean Burger with Barbecue Sauce
with New England IPA (BrewDog Hazy Jane)

SAUCE 2: BLACK BEAN

Chinese Stir-fry with Black Bean Sauce
with Scotch Ale (Traquair House Ale)

SAUCE 3: BOLOGNESE

Spaghetti Bolognese
with Best Bitter (Harvey's Sussex Best)

HOLD THE SAUCE

To illustrate the dominance of sauce, cook a small batch of each dish without it. For example, a plain bean burger, plain chicken or plain minced meat. Pair them with the beers. Do they work as well? Kind of, but with an empty feeling served on the side.

SPICY BEAN BURGER WITH BARBECUE SAUCE & NEW ENGLAND IPA (BREWDOG HAZY JANE)

COURSE 1: SAUCE

BARBECUE SAUCE

Here's the caveat: barbecue sauce means many things to many people, or – more accurately – one different thing to many people. Whether you drench your meat of choice in a sauce majoring in mustard, tomatoes, molasses, or vinegar, pairing Barbecue with a beer can be a tricky proposition. Lager or pale ale works with grilled meat, but the ferocity, rich depth, or tart prickle (delete as applicable) that elevates your favorite barbecue sauce tends to outfox these beers. Thankfully, modern craft brewing has your back. Over the last few years, a style has arrived that more than works with barbecue sauce. And the key is its softness.

Open up our hazy heroine and pour, checking just how opaque beer can be when brewers set their minds to it. These hazebombs are par for the course these days, as breweries opt not to add finings or filter their beers so that all of the flavor that lies in suspension is retained. The styles given this treatment are characteristically pale ales or IPAs and these hazy beers are softer, fruitier, and have a smoothness of delivery that makes them hugely easy to drink, whatever their ABV.

SERVES 2

Ingredients:

14oz (400g) can black beans, drained
14oz (400g) can kidney beans, drained
vegetable oil, for frying
1 shallot, finely sliced
1 garlic clove, finely chopped
½ teaspoon ground cumin
½ teaspoon smoked paprika
½ teaspoon mild chili powder
salt and pepper, to taste

To serve:

2 burger buns, halved
Barbecue sauce of your choosing
mayonnaise
your pick of toppings (pickles, lettuce, tomato, etc.)

Dry the drained beans with paper towel to remove as much moisture as possible. Put the beans into a large mixing bowl and crush them with a fork (or a potato masher). This is the start of your burger mix.

Heat a drizzle of oil in a frying pan over a medium heat and fry the shallot and garlic until browned. Add the spices and stir through while still warm, then allow to cool.

Add the spiced mixture to the crushed beans and stir to combine. Season to taste. Form the mixture into 2 burger-shaped patties.

Shallow-fry the patties in 1 tablespoon of vegetable oil until browned on both sides, turning carefully with a spatula. Alternatively, these burgers can be baked in the oven at 400°F/200°C for 30 minutes – frying is faster but can blacken the outside, which may or may not be to your liking. As burgers are ex-grill food and now fast-food, how you cook them relates to how much time you have. If you are frying, drop the halved burger buns into the remnants of the oil to toast them a moment before compiling your burger.

Before building your burgers, add a generous blob of mayo to one half of the bun and Barbecue sauce to the other half. Fill your burgers as you wish. Heaven awaits.

PAIRING NOTES:

New England IPA majors in citrus and tropical flavors, without the biting bitterness of other IPAs. The juiciness acts to counter the dry spiciness of the burger, but more importantly the carbonation and body cuts through the sweeter Barbecue sauces while acting in perfect opposition to the sharper, more vinegary ones. It's such a team player – whatever your sauce of choice – and that maybe explains why hazy IPA has entered the craft beer market in such a huge way.

OTHER BEERS TO TRY:

- Sierra Nevada Hazy Little Thing
- Verdant Bloom
- Thornbridge Green Mountain

CHINESE STIR-FRY WITH BLACK BEAN SAUCE & SCOTCH ALE (TRAQUAIR HOUSE ALE)

COURSE 2: SAUCE
BLACK BEAN SAUCE

Let's start with the beer here. The brewery at Traquair House is one of the UK's oldest family-owned breweries, dating to 1965 but with a lineage from the early 1700s. Boiled in a copper dating to 1738 and fermented in oak vessels that are over 200 years old, their beers are to Scotland what Cantillon is to Belgium. Chinese food has probably never been on their radar, but this marriage of flavors, of dark, bitter, umami-rich beans and Scotch Ale, is unbeatable. The traditional, heavy beers of Scotland are amazing with dark Asian sauces.

Open the Scotch Ale and savor its enormous, complex bouquet of aromas. There's a deep caramel, toffee edge that leads into a woodsy, tree-bark note (likely from the time spent in oak at the brewery). Sweet dried fruit also takes centerstage. It's a fantastic beer to savour, in your deepest armchair, but also one that is ideal to cook with. So let's fire up the wok and give it something to work on.

You can make your own black bean sauce if you can get your hands on fermented black beans – there are numerous ways online in which to do it (you can also use canned black beans if you have a lot of firepower to add to them). Pre-made black bean sauce is a bit of a cop-out, but it guarantees the balance of rich umami you're looking for.

SERVES 2

Ingredients:

2 tablespoons vegetable oil
1 red pepper, deseeded and thinly sliced
1 onion, thinly sliced
2 garlic cloves, finely chopped
1¼-in (3-cm) piece of fresh root ginger, peeled and finely chopped
1 green chili, deseeded and finely sliced
2 bok choy, leaves separated
2 chicken breasts, thinly sliced
3 tablespoons black bean sauce
2fl oz (50ml) chicken stock
3 spring onions, sliced
large bunch of cilantro, leaves picked
6oz (175g) cooked white rice, to serve

Heat the oil in a large wok over a high heat until hot and start by quick-frying the pepper and onion. Add the garlic, ginger, and chili, and then the bok choy. The bok choy will take only a moment to wilt – you want the white part near the knuckle to have a little bite to it. Throw in the chicken and stir-fry until cooked through, then it's sauce time.

Add the black bean sauce to the wok along with the stock and stir around until the stir-fry is well coated with sauce. Just before serving, stir in the spring onions and cilantro.

Serve immediately with the rice and the Scotch Ale.

PAIRING NOTES:

The key to this pairing is umami. The newest sensation on the taste block is there in spades in the tangy, bitter, and slightly funky beans – and matched by the roasted and chocolate-rich malts in the beer. That sweet raisin flavor we mentioned is also the perfect foil for the warmth of the beans and the heat of the stir-fry.

OTHER BEERS TO TRY:

- Tempest Old Parochial
- Oskar Blues Old Chub
- Gordon Scotch Ale

SPAGHETTI BOLOGNESE & BEST BITTER (HARVEY'S SUSSEX BEST)

COURSE 3: SAUCE

BOLOGNESE SAUCE

It's one of the world's most-used ingredients and yet one of the hardest with which to initiate a beer pairing. *Solanum lycopersicum*, a.k.a. the humble tomato, is a tough sell when rummaging through your beer cupboard. Used as a vegetable, tomatoes are actually a fruit but classified as a berry – so maybe this confusion is to blame. But really, tomatoes are tricky because of their acidity and umami qualities; they can make bitter beers harshly astringent. Also, one of beer's main off-flavors (DMS) smells like tomato sauce. So, where to turn? As ever, with most questions, the answer is "the pub." Best Bitter buddies up to tomatoes.

Your bitter and your pint glass await. These are easy-drinking, go-to British ales designed to be followed one after the other, but we are following this one with the most famed Italian dish that doesn't really exist in its homeland. The city of Bologna gave the world tagliatelle, tortellini, and lasagne – but not the dish named after it. Various beef and tomato "ragù" exist in parts of Italy, but very few authentic ones are made with ground beef, and zero are ladled on top of spaghetti. They are also never served alongside British bitter. But we are here to prove they can be.

SERVES 2

Ingredients:

1 tablespoon vegetable oil
1 onion, finely diced
1 carrot, finely diced
2 garlic cloves
9oz (250g) ground beef
14oz (400g) can chopped tomatoes
1 tablespoon tomato purée
1 bay leaf
1 teaspoon dried oregano
1 teaspoon mixed herbs
2fl oz (50ml) red wine (don't @ us)
2fl oz (50ml) beef stock
4½oz (125g) spaghetti
salt and pepper to taste

Heat the oil in a large frying pan over medium heat, add the onion and carrot and fry until softened, then add the garlic and cook for about 30 seconds until fragrant. Push to one side of the pan, add the ground beef and cook until browned all over. Stir into the onion and carrot until well combined, then you can add all of the other ingredients in turn and slowly bubble away – how long is up to you. Season to taste before serving. If the sauce reduces too much you can add a little water to bring it back, but cook until it is the consistency you're looking for.

Cook the spaghetti according to the packet instructions and serve alongside the Bolognese sauce. Just don't tell *nonna*.

PAIRING NOTES:

You may not think it, given the name of the style, but Best Bitters work here because of their sweetness. The key to taming the tomato is that very factor; it balances the acidity of the tomato and keeps things from becoming unpleasantly bitter (another irony). There's also a secondary pairing in action with the biscuity, slightly roasty malt and the browned minced meat, so – as with all good pairings – everything works out in the end.

OTHER BEERS TO TRY:

- Coniston Bluebird Bitter
- Adnam's Southwold Bitter
- Timothy Taylor Boltmaker

PIZZA

New Haven, Connecticut, is home to much more than Yalies. In this backyard of Ivy-League superiority, there also lies a superior form of the Italian dish that has taken over the world. Pizza has pushed into kitchens everywhere, but in this corner of Connecticut, the addition of a single letter has elevated it to a whole other meaning and something approaching a way of life. This is an "apizza" town.

A Neapolitan dialect gave the dish its second A and families immigrating to the United States brought it to Connecticut in droves. Throughout this book we have talked about the craft beer community, but that's got nothing on the New Haven apizza community. These thin, crispy oval pizzas with charred edges holding up unique toppings have achieved legendary status.

Is this deserved? There's only one way to find out. If you find yourself bored with Broadway, then head less than two hours up the coast from New York for an entirely new outlook on the world's favorite comfort fast food. What started out as original tomato pies, topped without cheese (you have to ask for it, even today), have become a series of truly unique classics – including Pepe's white clam pie.

Pepe's was founded by Frank Pepe in 1925 and they have made millions of their signature apizzas, topped with half a dozen ingredients: romano cheese, garlic, olive oil, parsley and clams. The littlenecks are sweet, chewy, and give a briny flavor that acts as another seasoning to this herbal, doughy superstar. There's a reason this has been named the number one pizza in America – and not just in alphabetical order. Apizza is it.

BEER PAIRING

Pilsner. Bready dishes are best served paired with pale, bready Pilsners; the balance of malt and dough makes everything work beautifully. The assertive bitterness of the Pils cuts through the cheesy goodness and the sweetness of the clams, while the herbs and punchy garlic act to keep the beer honest and never let it overpower the multitude of flavors on your plate. Apizza and a Pilsner, *per favore*.

CHEESE

MATCH OR CONTRAST THEME: CHEESEATHLON

A three-stage cheese meal can only really be called a Cheeseathlon, and we have a real humdinger for you here. To end our run of amazing pairings and matchings, wheel out the trolley of wonder and reach for those little strange-shaped knives. We've all experienced a cheese board to end a meal (or bypassed it for dessert), and typically these come in restaurants with port, sherry, or some other fortified wine served alongside.

There is an alternative. Why have a few cheeses with a single drink, when you can have each cheese paired to its own?

Beer and cheese have been bedfellows for centuries, enjoyed in everything from dark, dusty roadside taverns to the dazzling French countryside. You can go for versatile cheeses that work with a number of different beers, or vice versa, or you can select individual notes and tune into those instead. We, as you may have gathered, are doing the latter. If you haven't tried beer and cheese, then you are in for a treat.

The principles are the same as with other stages of a menu – you can match complementary flavors in your cheese and beer, or you can select one that offsets the other in a contrasting way and brings both to life. There are nearly as many flavors and varieties of cheese as there are types of malt, strains of hops, and flavors derived from yeast – so you can really go to town to end this pairing adventure.

CHEESEATHLON

LEG 1: SOFT-RIPENED WITH BELGIAN BLONDE

Baked Camembert
paired with brasserie De La Senne ZinneBir

LEG 2: SMOKED WITH FESTBIER

Smoked Cheese Empanada
paired with Augustiner Edelstoff

LEG 3: STRONG BLUE WITH IMPERIAL STOUT

Stilton Rarebit with Walnuts
paired with Great Divide Yeti

CHEESE DREAMS

You may have noticed that all three cheeses and all three beers here are full-on, full-flavored. You could match a mild Cheddar to an amber ale, or a waxy European breakfast cheese to a light lager – but this is the end of the meal. Pair accordingly.

BAKED CAMEMBERT & BRASSERIE DE LA SENNE ZINNEBIR

LEG 1: CHEESEATHLON
SOFT-RIPENED WITH BELGIAN BLONDE

If you've been lucky enough to drink Belgian beer in its birthplace, chances are you may have been served up a small bowl of cheese – maybe perched on a lace doily – with your bottle or draft of choice. This *fromage* or *kaas* (depending on what region of Belgium you are in) is just perfect with the local beer, so any meal that espouses the benefits of beer and cheese has to feature a Belgian offering. Its blonde ales are perfect openers here; highly carbonated, with a head of pillowy foam and shining like a noontime sun over a barley field, they are ideal when enjoyed alongside a soft, bloomy rind cheese such as Camembert. And this recipe literally couldn't be easier.

Open up your Belgian Blonde and pour into a large, stemmed glass. This is a beer that comes to life as you watch, with the soft rising trails of carbonation becoming a giant, pillowy foam that lifts up and sometimes out of the glass. The best verb to describe these beers as you drink them is "tingling" – they are effervescent, dry, and fruity but not as full-forward as their second cousins, the saisons. Anyway, all of this discovery can take place as you preheat the oven to 350°F/180°C and reach for the cheese.

It is best to bake Camembert when the cheese is at room temperature, so resist the temptation to keep it in the fridge. This recipe is one of the rare ones where all the prep – and it's so basic it's not even really prep – is done beforehand, so once the cheese is in, you can put your feet up until it comes out. There's one golden rule to follow: keep the cheese in its box. Otherwise, you'll have a puddle. For best results, wrap the circular wooden box in foil to keep the gooey cheese inside.

Toss some sprigs of rosemary onto the top of the cheese, then drizzle it with olive oil. Season generously and bake in the oven for 15 minutes, until it's molten and ready to go.

Remove and eat as soon as you can stand it, with the bread for dipping. Toast also works here. Eating has never come easier, or more rewarding.

SERVES 4

Ingredients:

9oz (250g) Camembert cheese (boxed), at room temperature

handful of rosemary sprigs

olive oil, for drizzling

salt and pepper

crusty bread, to serve

PAIRING NOTES:

This pairing works because neither thing overpowers the other. The slight herbal edge from the rosemary works with the spiciness of the Belgian yeast, and the ivory-colored, creamy cheese is cut by the carbonation and the hop-derived bitterness. Bloomy cheese has a slight lactic savory tang to it, from the rind, and this acidity is contrasted by the alcohol-derived richness and warmth from the beer.

OTHER BEERS TO TRY:

- La Trappe Blond
- Petrus Blond
- Allagash Blonde

OTHER CHEESES TO TRY:

- Brie de Meaux (or any Brie, really)
- Reblochon
- Goats' cheese

ZINNEBIR

SMOKED CHEESE EMPANADA & AUGUSTINER EDELSTOFF

LEG 2: CHEESEATHLON
SMOKED WITH FESTBIER

From one heartland of European brewing to another. The bright, full-bodied lagers of southern Germany are more than just beer-tent fodder and deserve more time in the limelight than the three autumnal weeks they are given. From Oktoberfest to the world, Festbier is smooth, golden, and distinctly more hop-forward than your typical helles. As such, Festbier is ideal for the cheese that is more flavorful than your typical cheese. You could (and should) pair smoked cheese with rauchbier, but Festbier is arguably even better. And if that's an argument to be had, it's one for shouting across a long, noisy table.

This recipe calls for at least a few bottles of Augustiner's cream-and-gold-labeled Festbier, maybe pulled right from one of their iconic blue crates, for instance. More than any other, this style is at its best when taken one after the other in your favorite company – it was what the Festbier was specifically designed for. Export-strength Helles is a mighty beer, set for a party, so let's give it one with South American-inspired finger food. Before that, open up a bottle and check out the voluminous, rocky-foamed pour you get. Savor those aromas of earthy spice, grassy floral hop notes, and the bready, crusty malts.

It's time to make pastry. Find a large bowl and add the flour, baking powder, and salt and give them a stir. Take the butter from the fridge and cut it into small cubes before mixing into the same bowl by hand until it looks like bread crumbs. Pour in a small amount of water, just enough to let you squeeze the mixture into a ball of dough, then cover in plastic wrap and refrigerate for at least 1 hour.

Meanwhile, enjoy your Edelstoff and heat 1 tablespoon of the oil in a frying pan over medium heat. Add the onion and fry until translucent, then transfer onto paper towel to soak up the cooking oil and dry the slivers out. Grate the smoked cheese into a bowl and, when the onion is cold, mix together.

Take your pastry out of the fridge and pull it into around 12 equal-sized balls, before rolling each one out to about 1⁄16in (2mm) thick. Use a coffee cup as a template and cut around it to make each circular empanada case the same size. Fill each circle of dough with a small handful of the smoked cheese and onion mixture, then brush the edge of the dough circle with beaten egg and carefully seal it shut into a half-moon-shaped pielet. Crimp the edges with a fork to keep the cheese where it should be.

Heat another 2–3 tablespoons of oil in a deep frying pan and shallow-fry the empanadas in batches until golden brown on each side. Drain on paper towel and eat when still warm and gooey.

SERVES 4

Ingredients:

For the pastry:

1lb 2oz (500g) plain flour
2 teaspoons baking powder
1 teaspoon salt
5oz (150g) cold unsalted butter, cubed

For the filling:

3–4 tablespoons vegetable oil
1 onion, finely sliced
14oz (400g) smoked cheese, grated
1 egg, beaten

PAIRING NOTES:

Warm, smoked cheese is hard to beat, and this pairing works because of the malt backbone in the beer. The bready, crusty notes are mirrored in the crispy pastry, and then the smoke from the cheese arrives to complement the malt a second time. The beer's dryness helps with the fatty cheese and the oily onions, with the carbonation cleaning the palate and leading to another empanada arriving shortly.

OTHER BEERS TO TRY:

- Weihenstephaner Festbier
- Hofbräu Festbier
- Victory Festbier

OTHER CHEESES TO TRY:

- Gruyère
- Manchego
- Comté

YETI
Imperial Stout

STILTON RAREBIT WITH WALNUTS & GREAT DIVIDE YETI

LEG 3: CHEESEATHLON
STRONG BLUE WITH IMPERIAL STOUT

Is there anything more comforting than cheese on toast? Try cheese on toast and imperial stout. The most reassuring, enveloping beer you can have in your cupboard is a hug in a snifter. Anything that warms you as much as sitting beside the fire is a welcome way to end a meal, and rich, boozy dark stout is that thing. This final act in our beer and food odyssey is lifted by the addition of walnuts, to yield a second level of flavor matching – a fitting end, and proof that beer is a natural partner to both high-end dining and the humble grilled cheese.

It's Yeti time. This is a beer that the Great Divide Brewing Company describes as an "onslaught of the senses" (in a good way) and is right up there with the best imperial stouts around. Big, bold, and bruising, it is as rewarding as beer can get. Open it up and drink it in. Caramel, toffee, and darkly roasted malt at first, but then something not often found in thick, black beers – a hit of American hops. Yeti is more turbo Black IPA than imperial stout and is a beer that defies being pigeonholed, as all the best do.

To give the beer a royal welcome, first turn on the grill to preheat.

Throw the Stilton into a bowl and crumble up with a fork. Add the chopped walnuts to the bowl, then drop in the egg yolk and a dash of Yeti from the glass (as much as you can spare), season with salt and pepper and mix with the fork. Put the bread under the grill and toast on one side – keep an eye on it, as burnt rarebit is a disaster. Once browned to your liking, remove and flip so that the ungrilled side is face up. Spread the cheese mixture onto the bread and return to the grill to brown until bubbling hot and irresistible.

SERVES 4

Ingredients:

9oz (250g) Stilton cheese
1¾oz (50g) walnuts, chopped
1 egg yolk
a slug of Great Divide Yeti
4 thick-cut slices of bread
salt and pepper

PAIRING NOTES:

The roasts have it – the toasted bread, the savory nuts, the malt-led stout. Putting the beer in the topping brings all the elements together, and if your bread of choice has a dark crust, then it will work all the better with the Yeti. Stilton is a cheese that goes toe-to-toe with the biggest hitters: port, brandy, barley wine – but also, in a worthy addition to this pantheon, imperial stout.

OTHER BEERS TO TRY:

- Bell's Expedition
- Thornbridge Saint Petersburg
- North Coast Old Rasputin

OTHER CHEESES TO TRY:

- Roquefort
- Gorgonzola
- Maytag Blue

DIY DOG

BREWDOG
BREWDOG
BREWDOG

DIY DOG

GIVING IT ALL AWAY. AGAIN AND AGAIN.

Home-brewing inspires and launches the careers of brewers around the world, and has done so for decades. It is the foundation of our entire industry. Almost all of our brewhouse team started their interest in brewing by working on their own creations, and even today we still use a 13¼-gallon (50-liter) home-brew system to develop new beers and formulate new recipes. The spirit of home-brew runs through our DNA.

The reason why is a simple one – even with the smallest 1-gallon home-brew kit, you can make amazing beer. Our craft is entirely and eminently scalable down to even the smallest degree. If you know the basics, are prepared to learn as you go, and strive to keep your equipment and kit clean, you can make beer as good as we do in your own kitchen. You can have your own brewery in your garage.

One of the greatest resources out there for anybody looking to create their own beer is the internet. Online, you'll find a wealth of information, advice, and ways to ace your home-brew – there is so much out there for fledgling brewers that you can find the answer to any question, learn how to right (almost) any wrong, and be tempted into the next of many directions for your home-brew journey.

Looking at all of this, from where we ended up, led us to thinking that we could play our part too. Having benefitted from the support of the global home-brew community when we started out in the late noughties, a decade after that we were determined to give something back. We wanted to pay tribute to our home-brewing roots. So, in 2016, we gave away every recipe we had ever made, for free, online.

Just search for "DIY Dog" to download our entire back catalog.

DIY Dog was the best thing we have ever done. More than ten years in the making, it now contains over 400 BrewDog recipes to re-create at home. Its most recent update runs from the beer that started it all for us – original-recipe Punk IPA – to recipe #415: Aplomb Bomb, a 7.0% ABV plum and vanilla Scottish sour from BrewDog OverWorks. More will follow, every year that passes from the moment you first read this.

As a liquid anthology, it is a thank you to everyone who has followed our adventures and had a go at brewing our beers in order to better understand our company. DIY Dog is the key to our kingdom and the recipes are there to be copied, amended, updated, and bastardized. In the same spirit, we published some of those recipes in our first book, *Craft Beer for the People,* alongside a chapter on how to home-brew.

DIY DOG IS THE KEY TO OUR KINGDOM.

And now, in this second book, they've returned. We got such a buzz from including them, we are doing it all again with more in the following pages. And, like the first book, they are followed by further recipes from other craft breweries we approached, many of which have never been released before. So fire up that brewkit and give it a try. It's what all of us would want.

PUNK IPA (2007–2010)

INDIA PALE ALE

FIRST BREWED 2007

ABV 6.0% **IBU 60** **OG 1.056**

BREWDOG ELLON, UK

Our flagship beer that kick-started everything. This is James and Martin's original take on an American IPA, subverted with punchy New Zealand hops that create an all-out riot of grapefruit, pineapple, and lychee before a spiky, mouth-puckering bitter finish.

BASICS

Volume	20L	5.3gal
Boil Volume	25L	6.6gal
ABV		6.0%
Target FG		1.010
Target OG		1.056
EBC		17
SRM		8.5
pH		4.4
Attenuation level	82%	

METHOD/TIMINGS

MASH TEMP

65°C 149°F 75 mins

FERMENTATION

19–21°C 66–70°F

BREWER'S TIP

While it may surprise you, this version of Punk IPA isn't dry-hopped but still packs a punch! To make the best of the aroma hops, make sure they are fully submerged and add them just before knock out for an intense hop hit.

FOOD PAIRING

Spicy carne asada with a pico de gallo sauce

INGREDIENTS

MALT

Extra Pale Malt	5.3kg	11.7lb

HOPS

Variety	(g/oz)	Add	Attribute
Ahtanum	17.5/0.62	Start	Bittering
Chinook	15/0.53	Start	Bittering
Crystal	17.5/0.62	Middle	Flavor
Chinook	17.5/0.62	Middle	Flavor
Ahtanum	17.5/0.62	End	Flavor
Chinook	27.5/0.97	End	Flavor
Crystal	17.5/0.62	End	Flavor
Motueka	17.5/0.62	End	Flavor

YEAST

Wyeast 1056 – American Ale™

PUNK IPA (2010–PRESENT)

INDIA PALE ALE

FIRST BREWED 2010

ABV 5.6% **IBU 40** **OG 1.053**

BREWDOG ELLON, UK

In 2010, we finally got our paws on the equipment we needed to dry-hop our beers. We focused all our energy on dry-hopping, amping up the aroma and flavor of our flagship beer to create a relentless explosion of tropical fruits, and adding a hint of Caramalt to balance out the insane amount of hops.

BASICS

Volume	20L	5.3gal
Boil Volume	25L	6.6gal
ABV		5.6%
Target FG		1.011
Target OG		1.053
EBC		15
SRM		7.6
pH		4.4
Attenuation level		78%

METHOD/TIMINGS

MASH TEMP

66°C 151°F 75 mins

FERMENTATION

19–21°C 66–70°F

BREWER'S TIP

To get the best possible profile from the dry hops, we recommend dry-hopping post fermentation for 5 days. Dry hops should be added at cellar temperature. We find 57°F (14°C) results in the most aromatic dry-hop profile.

FOOD PAIRING

Shredded chicken tacos with a mango-chili-lime salsa

INGREDIENTS

MALT

Extra Pale Malt	4.38kg	9.7lb
Caramalt	0.25kg	0.55lb

HOPS

Variety	**(g/oz)**	**Add**	**Attribute**
Chinook	20/0.71	Start	Bittering
Ahtanum	12.5/0.44	Start	Bittering
Chinook	20/0.71	Middle	Flavor
Ahtanum	12.5/0.44	Middle	Flavor
Chinook	27.5/0.97	End	Flavor
Ahtanum	12.5/0.44	End	Flavor
Simcoe	12.5/0.44	End	Flavor
Nelson Sauvin	12.5/0.44	End	Flavor
Chinook	47.5/1.7	Dry Hop	Aroma
Ahtanum	37.5/1.3	Dry Hop	Aroma
Simcoe	37.5/1.3	Dry Hop	Aroma
Nelson Sauvin	20.0/0.71	Dry Hop	Aroma
Cascade	37.5/1.3	Dry Hop	Aroma
Amarillo	10/0.35	Dry Hop	Aroma

YEAST

Wyeast 1056 – American Ale™

HAZY JANE

NEW ENGLAND IPA

FIRST BREWED 2017

ABV 7.2% IBU 30 OG 1.065

BREWDOG ELLON, UK

A Vermont-style IPA with low background bitterness, loaded with juicy fruit character. Pine, stonefruit, mango, light resin, and hints of lime peel – this juicy IPA is low in bitterness, full-bodied and smooth, enhancing the soft, ripe-fruit flavors.

BASICS

Volume	20L	5.3gal
Boil Volume	25L	6.6gal
ABV		7.2%
Target FG		1.009
Target OG		1.065
EBC		15
SRM		8
pH		4.2
Attenuation level		86%

METHOD/TIMINGS

MASH TEMP

66°C	151°F	25 mins

FERMENTATION

21°C	70°F

BREWER'S TIP

This beer is supposed to be very cloudy. Don't be shy about getting a good bit of the hot break into the fermenter.

FOOD PAIRING

Goats' cheese bruschetta

INGREDIENTS

MALT

Pale Malt	3.96kg	8.7lb
Maris Otter	0.96kg	2.1lb
Wheat Malt	0.6kg	1.3lb
Flaked Oats	0.24kg	0.53lb

HOPS

Variety	(g/oz)	Add	Attribute
Chinook	1/0.04	Middle	Flavor
Chinook	20/0.71	End	Aroma
Amarillo	20/0.71	End	Aroma
Simcoe	20/0.71	End	Aroma
Citra	50/1.8	Dry Hop	Aroma
Simcoe	50/1.8	Dry Hop	Aroma
Amarillo	50/1.8	Dry Hop	Aroma
Mosaic	50/1.8	Dry Hop	Aroma

YEAST

Wyeast 1056 – American Ale™

CLOCKWORK TANGERINE

CITRUS SESSION IPA

FIRST BREWED 2017

ABV 4.5%   IBU 37   OG 1.047

BREWDOG   ELLON, UK

Winner of the 2017 Prototype Challenge, Clockwork Tangerine is a session-strength IPA, infused with tangerine. It first appeared as a seasonal release in 2018, but proved so popular it was elevated to our year-round line-up the year after.

BASICS

Volume	20L	5.3gal
Boil Volume	25L	6.6gal
ABV	4.5%	
Target FG	1.012	
Target OG	1.047	
EBC	20	
SRM	10	
pH	4.2	
Attenuation level		75%

METHOD/TIMINGS

MASH TEMP

66°C   151°F   45 mins

FERMENTATION

19°C   66°F

BREWER'S TIP

Try your own fruit series with this beer: grapefruit, passion fruit, raspberries, or Morello cherries would be great.

FOOD PAIRING

Fresh green salad with avocados and pomegranate seeds

INGREDIENTS

MALT

Pale Malt	3.42kg	7.5lb
Caramalt	0.432kg	0.95lb
Medium Crystal	0.09kg	0.2lb

HOPS

Variety	(g/oz)	Add	Attribute
Simcoe	6/0.21	Start	Bittering
Citra	2/0.07	Start	Bittering
Simcoe	2/0.07	Middle	Flavor
Citra	2/0.07	Middle	Flavor
Ahtanum	5/0.18	Middle	Flavor
Chinook	5/0.18	Middle	Flavor
Chinook	3/0.11	End	Aroma
Simcoe	18/0.63	End	Aroma
Citra	5/0.18	End	Aroma
Mosaic	8/0.28	End	Aroma
Citra	25/0.88	Dry Hop	Aroma
Mosaic	25/0.88	Dry Hop	Aroma
Simcoe	40/1.4	Dry Hop	Aroma
Ahtanum	15/0.53	Dry Hop	Aroma
Chinook	19/0.67	Dry Hop	Aroma
Amarillo	4/0.14	Dry Hop	Aroma
Cascade	15/0.53	Dry Hop	Aroma

BREWHOUSE ADDITION

Tangerine Extract   7.5g   0.26oz   Flame Out

YEAST

Wyeast 1056 – American Ale™

LOST LAGER

DRY-HOPPED PILSNER

FIRST BREWED 2018

ABV 4.7% **IBU 37** **OG1.042**

BREWDOG ELLON, UK

A Pilsner that combines the light, crisp, and clean lager profile provided by Weihenstephan's house yeast, with the vibrant citrus and stonefruit aromas associated with the German hop Saphir. This lager is easy-going but has subtle depths; toast, hints of spice, and a zesty lime marmalade character.

BASICS

Volume	20L	5.3gal
Boil Volume	25L	6.6gal
ABV	4.7%	
Target FG		1.006
Target OG		1.042
EBC		5
SRM		3
pH		4.4
Attenuation level		86%

METHOD/TIMINGS

MASH TEMP

65°C1	49°F	65 mins

FERMENTATION

11°C	52°F

BREWER'S TIP

Temperature control is extremely important for lager. Keep fermentation temperature around 50–54°F (10–12°C) and a maturation period of 3–4 weeks at 34.7–35.6°F (1.5–2°C) is ideal.

FOOD PAIRING

Sashimi

INGREDIENTS

MALT

Pilsner Malt	3.36kg	7.4lb
Carapils Malt	0.24kg	0.53lb

HOPS

Variety	(g/oz)	Add	Attribute
Hallertauer Taurus	8/0.28	Start	Bittering
Select Spalter	15/0.53	Middle	Flavor
Select Spalter	15/0.53	End	Aroma
Saphir	30/1.1	End	Aroma
Saphir	30/1.1	Dry Hop	Aroma

BREWHOUSE ADDITION

Amyloglucosidase	1g	0.04oz

YEAST

Saflager W-34/70

ZOMBIE CAKE

CHOCOLATE PRALINE PORTER

FIRST BREWED 2018
ABV 5.0% **IBU 25** **OG 1.062**

BREWDOG ELLON, UK

Dark forces are at work in this devilishly good praline porter. Toffee and chocolate unite and come a-knocking. Open up to layers of smooth roasty character, with notes of vanilla, mellow coffee, and a subtle nuttiness, and a bittersweet cliff-hanger finale.

BASICS

Volume	20L	5.3gal
Boil Volume	25L	6.6gal
ABV		5.0%
Target FG		1.020
Target OG		1.062
EBC		90
SRM		46
pH		4.2
Attenuation level		68%

METHOD/TIMINGS

MASH TEMP

65°C 149°F 45 mins

FERMENTATION

19°C 66°F

BREWER'S TIP

Add the honey close to the end of the whirlpool, to ensure the aroma stays in the beer.

FOOD PAIRING

Almond milk pudding

INGREDIENTS

MALT

Pale Malt	3.12kg	6.9lb
Caramalt	0.84kg	1.9lb
Simpson's T50	0.36kg	0.8lb
Brown Malt	0.48kg	1lb
Carafa Special Type 3	0.12kg	0.26lb

HOPS

Variety	(g/oz)	Add	Attribute
Amarillo	6/0.21	Start	Bittering
Bramling Cross	20/0.71	Middle	Flavor
Amarillo	4/0.14	Middle	Flavor

BREWHOUSE ADDITION

Vanilla Extract	65g	2.3oz	FV Addition
Milk Sugars (Lactose)	360g	12.7oz	Start of boil
Honey	240g	8.5oz	Whirlpool

YEAST

Wyeast 1272 – American Ale II

PUMP ACTION POET

STONE FRUIT IPA

FIRST BREWED 2017

ABV 7.5% **IBU 40** **OG 1.070**

BREWDOG ELLON, UK

A small-batch canned release that combines the sweet and juicy flavor of stonefruit with tropical hop aromas. Pump Action Poet is a riot of rich, welcoming fruit and pithy zest – and looks incredible in custom artwork from Tracie Ching.

BASICS

Volume	20L	5.3gal
Boil Volume	25L	6.6gal
ABV		7.5%
Target FG		1.013
Target OG		1.070
EBC		12
SRM		6
pH		5.2
Attenuation level		81%

METHOD/TIMINGS

MASH TEMP

66°C 151°F 40 mins

FERMENTATION

19°C 66°F

BREWER'S TIP

Adding the fruit at the end of the boil effectively pasteurizes it. If you plan to add fruit after primary fermentation, buy aseptic purée or have a plan to pasteurize it before adding it.

FOOD PAIRING

Spicy chili prawns

INGREDIENTS

MALT

Pale Malt	4.86kg	10.7lb
Wheat Malt	0.54kg	1.2lb
Flaked Oats	0.54kg	1.2lb

HOPS

Variety	(g/oz)	Add	Attribute
Amarillo	40/1.4	Start	Aroma
Simcoe	40/1.4	End	Aroma
Citra	30/1.1	Dry Hop	Aroma
Simcoe	30/1.1	Dry Hop	Aroma
Mosaic	30/1.1	Dry Hop	Aroma

BREWHOUSE ADDITION

Apricot Juice	1kg	2.2lb	Flame Out
Peach Juice	1kg	2.2lb	Flame Out

YEAST

Wyeast 1056 – American Ale™

BLACK EYED KING IMP (VIETNAMESE COFFEE EDITION)

IMPERIAL COFFEE STOUT

FIRST BREWED 2014

ABV 12.7 **IBU 85** **OG1.113**

BREWDOG ELLON, UK

A coffee edition of a stout first brewed in 2012. At 12.7% ABV, it is a super-intense and twistedly complex brew, with intense notes of sweet vanilla, rich espresso, smooth molasses, and bitter chocolate, barely contained by the container it's in.

BASICS

Volume	20L	5.3gal
Boil Volume	25L	6.6gal
ABV		12.7%
Target FG		1.038
Target OG		1.113
EBC		250
SRM		125
pH		5.2
Attenuation level		76.8%

METHOD/TIMINGS

MASH TEMP

65°C 149°F 50 mins

FERMENTATION

18°C 64°F

BREWER'S TIP

Buying top-notch coffee beans make a huge difference here. Give them a very coarse grind to get the most out of them.

FOOD PAIRING

Twenty-hour smoked brisket

INGREDIENTS

MALT

Extra Pale	6.25kg	13.8lb
Wheat	1.25kg	2.75lb
Caramalt	1.25kg	2.75lb
Crystal	1.56kg	3.4lb
Dark Crystal	0.625kg	1.4lb
Amber	0.625kg	1.4lb
Brown	0.625kg	1.4lb
Chocolate	0.625kg	1.4lb
Roasted Barley	0.312kg	0.69lb

HOPS

Variety	(g/oz)	Add	Attribute
Magnum	62.5/2.2	Start	Bittering
Willamette	31.25/1.1	End	Bittering
First Gold	31.25/1.1	End	Aroma/ Bittering

BREWHOUSE ADDITION

Coffee Beans	12.5g	0.44oz	End of Boil
Milk Sugars (Lactose)	125g	4.4oz	FV

YEAST

Wyeast 1272–American Ale II

FUNK × PUNK

BRETT-FERMENTED INDIA PALE ALE

FIRST BREWED 2018

ABV 5.5% **IBU 42** **OG 1.058**

BREWDOG OVERWORKS ELLON, UK

Funk x Punk takes our passion for hops and our infatuation with wild flavors to a natural conclusion – the juicy, tropical notes from our signature hopload marry together with the dry, funky Brett addition in perfect harmony. Enjoy now or cellar for years to come.

BASICS

Volume	20L	5.3gal
Boil Volume	25L	6.6gal
ABV	5.5%	
Target FG	1.004	
Target OG	1.058	
EBC	24	

METHOD/TIMINGS

MASH TEMP

65°C	149°F	55 mins

FERMENTATION

20°C	68°F

BREWER'S TIP

Rack the beer off the hops after around 4 days. This will help to avoid too much raw hop flavor in your finished beer.

FOOD PAIRING

Belgian mussels and frites

INGREDIENTS

MALT

Weyermann Pils	1.6kg	3.5lb
Weyermann Pale Wheat	1.2kg	2.6lb
Rye	0.5kg	1.1lb

HOPS

Variety	(g/oz)	Add	Attribute
Hallertauer Taurus	6/0.21	Start	Bittering
Simcoe	10/0.35	Flame Out	Bittering
Cascade	10/0.35	Flame Out	Bittering
Hallertauer Blanc	10/0.35	Flame Out	Bittering
Simcoe	20/0.71	Dry Hop	Aroma
Citra	20/0.71	Dry Hop	Aroma
Mosaic	20/0.71	Dry Hop	Aroma

YEAST

House Brett Blend

COSMIC CRUSH TROPICAL

TROPICAL SOUR ALE

FIRST BREWED 2018

ABV 5.8% **IBU 15** **OG 1.058**

BREWDOG OVERWORKS **ELLON, UK**

Cosmic Crush is Overworks year-round-available range of sours that focus on single-fruit additions. Aside from this one, which contains a quartet of different tropical fruit! An underlying Brett funk and bone-dry attenuation completes this triumphantly tropical sour.

BASICS

Volume	20L	5.3gal
Boil Volume	25L	6.6gal
ABV	5.8%	
Target FG	1.004	
Target OG	1.058	
EBC	15	

METHOD/TIMINGS

MASH TEMP

65°C	149°F	55 mins

FERMENTATION

20°C	68°F

BREWER'S TIP

The tropical fruits can produce a lot of sulphur. A pitch of fresh yeast for a healthy re-fermentation should help deal with this.

FOOD PAIRING

Tacos al pastor

INGREDIENTS

MALT

Weyermann Pils	1.9kg	4.2lb
Wheat	1.2kg	2.6lb
Flaked Oats	0.3kg	0.7lb

HOPS

Variety	(g/oz)	Add	Attribute
Magnum	1.5/0.05	Start	Bittering
Hallertauer Blanc	7/0.25	Flame Out	Bittering
Hallertauer Mittelfrueh	4/0.14	Flame Out	Bittering

BREWHOUSE ADDITION

Guava purée	0.5kg	1.1lb	FV
Papaya purée	0.5kg	1.1lb	FV
Pineapple purée	0.5kg	1.1lb	FV
Mango purée	0.5kg	1.1lb	FV

YEAST

House Brett Blend

AB:24

COFFEE & BROWN SUGAR BALTIC PORTER

FIRST BREWED 2017

ABV 11.5% **IBU 50** **OG 1.110**

BREWDOG ELLON, UK

AB:24 is a huge, mocha-forward beer laden with the aromas of freshly brewed coffee, toasty caramel, red berry fruit, and a bitter chocolate undercurrent. It is designed to be rich, warming, and full-flavored, with an almost brandy-like backbone from the muscovado and double-digit ABV.

BASICS

Volume	20L	5.3gal
Boil Volume	25L	6.6gal
ABV		11.5%
Target FG		1.019
Target OG		1.110
EBC		220
SRM		112
pH		4.4
Attenuation level		83%

METHOD/TIMINGS

MASH TEMP

66°C	151°F	75 mins

FERMENTATION

21°C	70°F

BREWER'S TIP

Do not be tempted to substitute the muscovado sugar with white sugar. It lacks the molasses (black treacle) component that comes from only being partially refined.

FOOD PAIRING

S'mores with smoked marshmallows

INGREDIENTS

MALT

Pale Malt	6.48kg	14.3lb
Double Roast Crystal	0.72kg	1.6lb
Munich	1.44kg	00lb
Chocolate Wheat	0.72kg	00lb
Medium Crystal	0.48kg	3.2lb
Flaked Oats	0.72kg	1.6lb
Oat Husks	0.048kg	0.11lb

HOPS

Variety	(g/oz)	Add	Attribute
Magnum	14/0.49	Start	Bittering
Galena	40/1.4	End	Aroma
Centennial	10/0.35	End	Aroma

BREWHOUSE ADDITION

Muscovado sugar	900g	2lb	Kettle
Ground coffee	20g	0.7oz	Flame O
Cold-brew coffee	160ml	5.4fl oz	FV

YEAST

Wyeast 1272–American Ale II

DOG H

12TH ANNIVERSARY IMPERIAL STOUT

FIRST BREWED 2019

ABV 15.5% **IBU 90** **OG 1.120**

BREWDOG ELLON, UK

Dog H is our liquid milestone, brewed to celebrate our 12th anniversary. Captivating dark malts, Naga chilis, cacao, and coffee. And then cognac. Dog H keys into the Auld Alliance thanks to the influential embrace of French brandy. This is barrel-aging done right.

BASICS

Volume	20L	5.3gal
Boil Volume	25L	6.6gal
ABV	15.5%	
Target FG		1.020
Target OG		1.120
EBC		300
SRM		152
pH		4.2
Attenuation level		83%

METHOD/TIMINGS

MASH TEMP

64°C 147°F 115 mins

FERMENTATION

21°C 70°F

BREWER'S TIP

For this beer to be successful, a few key points need to be followed. The first is picking the correct yeast strain, one that is known to have high alcohol tolerance. The second is that starting the brew with all the sugar necessary to gain +12% beer subjects the yeast to high osmotic stress and will hinder fermentation, so it is best to start with wort just over 1.100. Finally, once fermentation has decreased and moving approximately 1SG point per 24 hours, it is time to start adding dextrose to achieve target ABV.

FOOD PAIRING

Roquefort cheese

INGREDIENTS

MALT

Pale Malt	7.68kg	16.9lb
Wheat Malt	0.96kg	2.1lb
Extra Dark Crystal	0.84kg	1.9lb
Chocolate Wheat	0.6kg	1.3lb
Brown	0.6kg	1.3lb
Rye Malt	0.48kg	1.1lb
Flaked Oat Malt	0.96kg	2.1lb

HOPS

Variety	(g/oz)	Add	Attribute
Simcoe	40/1.4	Start	Bittering
Fuggles	40/1.4	End	Bittering/Aroma

BREWHOUSE ADDITION

Dextrose	1.2kg	2.6lb	Dry Hop
Coffee	40g	1.4oz	Dry Hop
Cocoa Nibs	40g	1.4oz	Dry Hop
Habanero Powder	1g	0.04oz	Dry Hop
Cognac-Soaked Oak Chips	200g	7oz	FV

YEAST

Wyeast 1272–American Ale II

JARL

SESSION PALE ALE

ABV 3.8% IBU 40 OG1038

FYNE ALES CAIRNDOW, ARGYLL, UK

Jarl is an easy-drinking, beautifully balanced session ale, bursting with flavor. Fyne Ales was one of the first UK breweries to use Citra hops when Jarl arrived in 2010. It has since become their flagship best-seller and a showcase for not just Citra but modern British brewing.

BASICS

Volume	20L	5gal
Boil Volume	21.5L	5.6gal
ABV		3.8%
Target FG		1008.5
Target OG		1038.0
EBC		4
SRM		2

pH pre-boil: below 5.9; post-boil: 5–5.4

Attenuation level	80%

METHOD/TIMINGS

MASH TEMP
65°C 149°F 60 mins

BOIL TIME
60 mins

FERMENTATION
Pitch at 70°F/21°C;
allow to rise to 74°F/23.5°C;
cool to 10°C/50°F for racking

FOOD PAIRING

Fish or seafood (anything from Loch Fyne, for instance!)

INGREDIENTS

MALT

Extra Pale Ale	2.7kg	6lb
Torrefied Wheat	0.3kg	0.66lb

HOPS

Variety	(g/oz)	Add	Attribute
Citra	75g/2.6	Start	Bittering

BREWHOUSE ADDITION

None

YEAST

Scottish/English Ale yeast – nothing too estery

BROKEN DREAM

STOUT

ABV 6.7% **IBU 30** **OG 1.073**

SIREN CRAFT BREW FINCHAMPSTEAD, BERKSHIRE, UK

An indulgence of chocolate and speciality malts, Broken Dream is smooth, unctuous, and addictive. It's brewed with milk sugar for balance and mouthfeel, along with carefully selected espresso from London's Climpson & Sons. Awarded CAMRA Supreme Champion Beer of Britain 2018.

BASICS

Volume	20L	5gal

Boil Volume sparge to 25L/6.6gal; boil back to 20L/5.25gal

ABV	6.5%
Target FG	1.022
Target OG	1.076
EBC	200
SRM	102
pH	4.3
Attenuation level	70%

METHOD/TIMINGS

MASH TEMP

66.5°C 152°F 80 mins

BOIL TIME

75 mins

FERMENTATION

start at 66°F/19°C; raise to 72°F/22°C when the SG drops below 1.040

FOOD PAIRING

Slow-cooked lamb shawarma

INGREDIENTS

MALT

Maris Otter	4.9kg	10.8lb
Malted Oats	0.64kg	1.4lb
Brown Malt	0.53kg	1.17lb
Chocolate Malt	0.42kg	0.93lb
Smoked Malt	0.21kg	0.46lb
Cararoma	0.105kg	0.23lb
Roast Barley	0.105kg	0.23lb

HOPS

Variety	(g/oz)	Add	Attribute
Apollo	10.5/0.37	Start	Bittering

BREWHOUSE ADDITION

Coffee	7/0.25	Boil Start
Lactose	10.5/0.37	10 mins before Boil End
Coffee	14/0.49	Dry Hop remove after 48 hours

YEAST

Siren House Vermont (22 million cells/ml)

LORD NELSON

SAISON

ABV 6.8% **IBU 24** **OG 1.056**

ELUSIVE BREWING WOKINGHAM, BERKSHIRE, UK

A powerhouse single-hop saison, released annually. The beer leads with a blend of pale malts, leaving the yeast and hops to interplay and really showcase this elusive and wonderful Kiwi hop. Originally brewed in collaboration with Weird Beard, in a neat twist the first iteration used hops borrowed from BrewDog.

BASICS

Volume	20L	5gal
Boil Volume	22L	5.75gal
ABV		6.8%
Target FG		1.056
Target OG		1.004
EBC		13
SRM		6.6
pH		4.4
Attenuation level	93%	

METHOD/TIMINGS

MASH TEMP

66°C 151°F 60 mins

BOIL TIME

60 mins

FERMENTATION

Pitch at 72°F (22°C) and allow to free rise (no temperature control) until activity slows, then add the dry hops

FOOD PAIRING

This dry and aromatic beer is great with spicy Thai dishes, such as red curry, or creamy rich cheeses

INGREDIENTS

MALT

Pale	1.8kg	8.3lb
Lager	1.8kg	8.3lb
Vienna	0.9kg	2lb
Munich	0.45kg	1lb

HOPS

Variety	(g/oz)	Add	Attribute
Nelson Sauvin	5/0.18	Start	Bittering
Nelson Sauvin	15/0.53	after 10 mins	Flavor
Nelson Sauvin	25/0.88	after 5 mins	Flavor
Nelson Sauvin	55/1.9	End	Aroma
Nelson Sauvin	100/3.5	Dry Hop on day 7	Aroma

BREWHOUSE ADDITION

None

YEAST

Belle Saison or Wyeast 3711 (French Saison)

MILK SHAKE

MILK STOUT

ABV 5.6% IBU 18 OG 1070.6

WIPER AND TRUE BRISTOL, UK

Milk stouts are brewed with sugar from cow's milk to give the beer a sweet, creamy tone, and Wiper and True's Milk Shake is one of the best around. It combines copious chocolate malts laced with vanilla and cacao to create a luxurious, milkshake-rich, and satisfyingly dark beer.

BASICS

Volume	20L	5gal
Boil Volume	22L	5.75gal
ABV		5.6%

Target FG 1027.5 (with lactose)/ 1019.0 (without lactose)

Target OG 1070.6 (with lactose)/ 1062.1 (without lactose)

EBC		252
SRM		128
pH		4.9–5.3
Attenuation level	61%	

METHOD/TIMINGS

MASH TEMP

67°C 153°F 60 mins

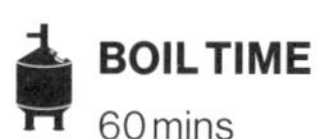

BOIL TIME

60 mins

FERMENTATION

20°C 68°F

FOOD PAIRING

Hearty foods, including roasted meats, sausages, and stews; the dark roasted malts also pair with rich, creamy cheeses and desserts such as chocolate mousse or ice cream

INGREDIENTS

MALT

GP	2.6kg	5.7lb
FMO	0.44kg	1lb
Wheat	0.24kg	0.5lb
Chocolate	0.24kg	0.5lb
PEC	0.12kg	0.3lb
Aromatic	0.12kg	0.3lb
GNO	0.08kg	0.2lb
Carafa III	0.08kg	0.2lb
Low-color Chocolate	0.04kg	0.09lb
Roasted Barley	0.04kg	0.09lb

HOPS

Variety	(g/oz)	Add	Attribute
Phoenix T90	7g/0.25oz	Start	Bittering

BREWHOUSE ADDITION

In the boil: lactose at 10 mins remaining, vanilla powder at and cocoa powder at end of the boil (ensure temperature is below 185°F/85°C upon addition of vanilla powder)

YEAST

BRY-97 (Lallemand)

EXPORT INDIA PORTER

PORTER

ABV 6% **IBU 45** **OG 1.060**

THE KERNEL LONDON, UK

The Kernel pride themselves on creating beers with lengths and depths of flavor, and this darkly resonant beer is a perfect example. A traditional London porter recipe lifted by an IPA-level hop load, it is one of the best dark beers available in the UK and a triumph of balancing flavor between malt and hops.

BASICS

Volume	20L	5gal
Boil Volume	23L	6gal
ABV	6%	
Target FG		1.015
Target OG		1.061
EBC		70
SRM		35

pH 4.2–4.4 in finished beer

Attenuation level	75%

METHOD/TIMINGS

MASH TEMP

66°C 151°F 60 mins

BOIL TIME

70 mins

FERMENTATION

68°F (20°C) rising to 73°F (23°C) at end of fermentation

FOOD PAIRING

Appleby's Cheshire or Kirkham's Lancashire cheeses

INGREDIENTS

MALT

Maris Otter Malt	3kg	6.6lb
Brown Malt	0.28kg	0.6lb
Chocolate Malt	0.28kg	0.6lb
Crystal Malt	0.28kg	0.6lb
Black Malt	0.14kg	0.3lb

HOPS

Note: The Kernel change the hops used in eac batch, but something like Columbus works really well, at the below rate of additions per liter. Bramling Cross also would work well, bu you would probably need to double the size of the additions as the AA% is much lower. O use a bittering hop with higher alpha values.

Variety	(g/oz)	Add	Attribute
See note	0.5/0.02	at 60 mins	Bittering
See note	0.5/0.02	at 15 mins	Flavor
See note	1/0.04	at 10 mins	Flavor
See note	1.5/0.05	at 5 mins	Flavor
See note	3/0.11	Dry Hop post-fermentation	Aroma

BREWHOUSE ADDITION

None

YEAST

American Ale yeast

VIBES HOPPY PILSNER

NEW WORLD HOPPY PILSNER

ABV 5.1% **IBU 35** **OG 1046**

ALMANAC BEER CO. ALAMEDA, CALIFORNIA

On the eastern side of San Francisco Bay is the city of Alameda, home to Almanac and their "farm to barrel" philosophy. Bold recipes, brewed sustainably, include this hybrid between a Czech-style lager and a modern pale ale. Ultra-crisp, with a big citrus aroma, it is a fantastic twist on the world's favorite beer style.

BASICS

Volume	20L	5gal
Boil Volume	21L	5.5gal
ABV		5.1%
Target FG		1008
Target OG		1046
EBC		6
SRM		3
pH		4.3
Attenuation level		83%

METHOD/TIMINGS

MASH TEMP

66°C 151°F 60 mins

BOIL TIME

60 mins (ensure whirlpool temperature is 169°F/76°C before hop recirculation)

FERMENTATION

55°F (12.5°C) for 12 days; dry-hop with 1 Plato remaining in fermentation

FOOD PAIRING

Ceviche

INGREDIENTS

MALT

Admiral Pilsner Malt	3.75kg	8.3lb
Weyermann Carafoam	0.35kg	0.77lb
Rahr Malted White Wheat	0.23kg	0.5lb

HOPS

Variety	(g/oz)	Add	Attribute
Huell Melon	8/0.28	First wort hopping	Bittering
Huell Melon	8/0.28	last 15 min of Boil	Flavor
Huell Melon	32/1.13	Whirlpool	Aroma
Motueka	16/0.56	Whirlpool	Aroma
Citra	16/0.56	Whirlpool	Aroma

BREWHOUSE ADDITION

None

YEAST

Omega Czech Lager II

DISCO SOLEIL

KUMQUAT IPA

ABV 6.5%　IBU 37　OG 1.062
(75% MASH EFFICIENCY)

BRASSERIE DIEU DU CIEL!　MONTREAL, CANADA

First brewed in 2013, this is the taste of liquid summer sunshine. Refreshing kumquat immerses the palate with citrus and tropical fruit, while biting hops interplay with bready malt. Whenever you find yourself yearning for those hazy summer days and warm electric nights, you will find solace with this spellbinding beer from Montreal's finest.

BASICS

Volume	20L	5gal
Boil Volume	22L	5.75gal
ABV		6.5%
Target FG		1.013
Target OG		1.062
EBC		10
SRM		5
pH		4.2
Attenuation level		78%

METHOD/TIMINGS

MASH TEMP

66°C　151°F　75 mins

BOIL TIME

60 mins

FERMENTATION

Pitch at 70°F (21°C) until complete attenuation. Hold for 2 days for diacetyl rest. Add the purée when the gravity reaches 1.019 (5° Plato)

FOOD PAIRING

Moderately spicy, wood-grilled jerk pork tenderloin

INGREDIENTS

MALT

Pilsner Malt	4.25kg	9.4lb
Carahell	0.5kg	1.1lb
Wheat Malt	0.5kg	1.1lb
Toasted Wheat Flakes	0.35kg	0.8lb

HOPS

Variety	(g/oz)	Add	Attribute
Apollo	3.5/0.12	First wort hopping	Bittering
Apollo	15/0.53	Start	Bittering
Citra	10/0.35	Middle	Flavor
Mandarina Bavaria	10/0.35	Middle	Flavor
Citra	20/0.71	End	Aroma
Mandarina Bavaria	20/0.71	End	Aroma

BREWHOUSE ADDITION

150g (5.5oz) kumquat purée at the end the fermentation

YEAST

Whitelabs (WLP007) Dry English Ale yeast

GO-TO IPA

SESSION IPA

ABV 4.8% IBU 65 OG 1.044

STONE BREWING SAN DIEGO, CALIFORNIA

Think of Stone Brewing's beers and you think of hops. Their pale ales and IPAs are heavy-duty hop bombs – even those that clock in at sub-5% ABV. Released in 2014, their session IPA Go-To doesn't shy away from the assertive character and alpha-led flavor profile we've come to expect. The key is extensive whirlpool hopping, which delivers the lupulin hit you look for – and always find – in California's finest. Another classic IPA from Stone, in a sessionable format.

BASICS

Volume	20L	5gal
Boil Volume	25L	6.6gal
ABV		4.8%
Target FG		1.011
Target OG		1.044
EBC		10
SRM		5
pH		4.5
Attenuation level	75%	

METHOD/TIMINGS

MASH TEMP

75°C 167°F rest for 30 mins

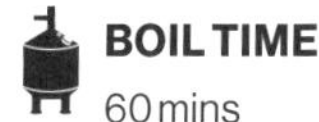

BOIL TIME

60 mins

FERMENTATION

20–22°C 68–72°F

FOOD PAIRING

Thai salad with peanut dressing, cucumber, and mint

INGREDIENTS

MALT

2-Row Pale	3.5kg	7.7lb
Caramel	0.12kg	0.26lb
Victory	0.12kg	0.26lb
Carapils	0.12kg	0.26lb

HOPS

Variety	(g/oz)	Add	Attribute
Chinook	69/2.4	last 10 mins of Boil	Aroma
Pekko	37/1.3	last 10 mins of Boil	Aroma
Cascade	32/1.13	last 10 mins of Boil	Aroma
Magnum	25/0.88	last 10 mins of Boil	Aroma
Mosaic	120/4.2	Whirlpool	Aroma
Amarillo	76/2.7	Whirlpool	Aroma
Pekko	63/2.2	Whirlpool	Aroma
Sterling	37/1.3	Whirlpool	Aroma
Crystal	37/1.3	Whirlpool	Aroma
Mosaic	210/7.4	Dry Hop (post fermentation)	Aroma
Citra	95/3.4	Dry Hop (post fermentation)	Aroma
Cascade	63/2.2	Dry Hop (post fermentation)	Aroma

BREWHOUSE ADDITION

None

YEAST

Whitelabs (WLP002) English Ale or (WLP007) Dry English Ale yeast

SUPERSONIC

DDH DIPA

ABV 8.5% **IBU 45** **OG 1.0852**

LERVIG STAVANGER, NORWAY

One of the most popular styles of the moment is the "juicy IPA." These smooth, hazy beers are often given double the dry-hop addition (DDH) – and when the beer itself is a Double IPA, you get a huge body with great drinkability. Mike Murphy, Brewmaster at Norway's Lervig Aktiebryggeri, is a master at this style, and Supersonic – its recipe appearing in print here for the first time – is, as he says, "a super tropical hoppy explosion that breaks the hopwave."

BASICS

Volume	20L	5gal
Boil Volume	28L	7.4gal
ABV		8.5%
Target FG		1.0852
Target OG		1.0218
EBC		16
SRM		8
pH		4.5

Attenuation level
62% (RDF)–73% (ADF)

METHOD/TIMINGS

MASH TEMP

68°C 154°F rest for 60 mins

BOIL TIME

90 mins

FERMENTATION

Pitch at 66°F (19°C), raise 1°F (0.5°C) per day to 71°F (21.5°C)

FOOD PAIRING

Grilled pineapple; habañero-and-honey-glazed chicken; sweet chili shrimp, tuna poké

INGREDIENTS

MALT

Golden Promise	6.25kg	13.8lb
Golden Naked Oats	0.32kg	0.7lb
Flaked Oats	0.9kg	2lb
Wheat Malt	0.62kg	1.4lb

HOPS

Variety	(g/oz)	Add	Attribute
No Bittering hop addition			
Amarillo	125/4.4	Whirlpool @ 176°F (80°C) with the wort under pH5.1	Flavor
Citra	156/5.5	Whirlpool @ 176°F (80°C) with the wort under pH5.1	Flavor
Simcoe	32/1.13	Whirlpool @176°F (80°C) with the wort under pH5.1	Flavor
Citra	78/00	Dry Hop (2 times) 24–36 hours into full fermentation	Flavor
Enigma	30/00	Dry Hop (2 times) 24–36 hours into full fermentation	Flavor
Simcoe	30/00	Dry Hop (2 times) 24–36 hours into full fermentation	Flavor

Repeat 1 day after active fermentation has cea
Remove yeast and hops after 3 days.

BREWHOUSE ADDITION

None

YEAST

Wyeast–London Ale III

TOPCUTTER IPA

INDIA PALE ALE

ABV 6.8% **IBU 70** **OG 1.0594**

BALE BREAKER YAKIMA, WASHINGTON

The epicenter of the hop industry in the Pacific Northwest is the city of Yakima, Washington. Over three-quarters of America's hops come through here, so hometown breweries are the envy of others around the world. If you brew an IPA in Yakima, you know it needs to be dry, bitter, and incredible. Bale Breaker's flagship Topcutter is all of these things and more.

BASICS

Volume	20L	5gal
Boil Volume	25L	6.6gal
ABV		6.8%
Target FG		1.0082
Target OG		1.0594
EBC		7.8
SRM		4
pH		4.5
Attenuation level	85.5%	

METHOD/TIMINGS

MASH TEMP

65.5°C 150°F 60 mins

BOIL TIME

75 mins

FERMENTATION

18–20°C 64–68°F

FOOD PAIRING

Roasted chicken, steak, Cheddar cheese

INGREDIENTS

MALT

2-Row Pale	5.84kg	12.9lb
Munich	0.16kg	0.35lb
Carapils	0.16kg	0.35lb
Caramel	0.14kg	0.3lb
Acidulated Malt	0.08kg	0.18lb

HOPS

Variety	(g/oz)	Add	Attribute
Warrior	15/0.53	Start	Bittering
Simcoe	15/0.53	last 15 mins of Boil	Flavor
Simcoe	20/0.71	End of Boil	Flavor
Citra	20/0.71	End of Boil	Flavor
Mosaic	20/0.71	End of Boil	Flavor
Loral	20/0.71	End of Boil	Flavor
Simcoe	60/2.1	Dry Hop	Aroma
Citra	80/2.8	Dry Hop	Aroma
Loral	60/2.1	Dry Hop	Aroma

BREWHOUSE ADDITION

None

YEAST

Wyeast 1056 – American Ale™

HAWT DIPA

DOUBLE IPA

ABV 8.0% IBU C.30 OG 1.0763

FUERST WIACEK BERLIN, GERMANY

The craft beer scene in Germany – in particular, Berlin – is changing almost by the day as new brewers arrive and new freedoms are discovered. The land of perfect lager is now also the land of perfect haze, and front and center in that regard are Fuerst Wiacek, Berlin's hazy magicians. Their Double IPA with Simcoe and Mosaic is a bonafide modern classic.

BASICS

Volume	20L	5gal
Boil Volume	25L	6.6gal
ABV		8.0%
Target FG		1.0157
Target OG		1.0763
EBC		8.5
SRM		4
pH		4.4
Attenuation level		78%

METHOD/TIMINGS

MASH TEMP

66°C 151°F rest for 60 mins

BOIL TIME

60 mins (8% boil-off)

FERMENTATION

Aerate properly, then ferment at 68°F (20°C), increase to 72°F (22°C) 48 hours into fermentation.

FOOD PAIRING

Thai red vegetable curry, smoky chicken wings

INGREDIENTS

Heidelberger Pilsner Malt	3.87kg	8.5lb
Wheat Malt	2.04kg	4.5lb
Carapils	0.34kg	0.75lb
Dextrose	0.34kg	0.75lb
Acidulated Malt	0.20kg	0.44lb

HOPS

Variety	(g/oz)	Add	Attribute
No Bittering hop addition.			
Columbus	100/3.5	Whirlpool	Flavor
Simcoe	200/7	End of fermentation	Aroma
Mosaic	200/7	End of fermentation	Aroma

BREWHOUSE ADDITION

None

Wyeast–London Ale III

TWO HEARTED ALE

AMERICAN IPA

ABV 7.0% **IBU 60** **OG 1.065**

BELL'S BREWERY COMSTOCK, MICHIGAN

One of America's most storied beers hails from Michigan's largest independent brewer – Bell's. Their classic US IPA Two Hearted was named best beer in America three years running (2017–2019) by the US Homebrewers Association, knocking Russian River's Pliny the Elder off the top spot. A single-hop Centennial IPA, Two Hearted is as good as American brewing gets.

BASICS

Volume	20L	5gal
Boil Volume	26L	6.9gal
ABV		7.0%
Target FG		1.011
Target OG		1.065
EBC		22.26
SRM		11.3
pH		5.2 (mash pH)
Attenuation level		83%

METHOD/TIMINGS

MASH TEMP

65.5°C 150°F 60 mins

BOIL TIME

60 mins

FERMENTATION

20–22°C 68–72°F

FOOD PAIRING

Barbecue brisket, caramel and apple cake, curried vegetable soup

INGREDIENTS

MALT

2-Row Pale	4.54kg	10lb
Pale Ale	1.36kg	3lb
Carapils	0.16kg	0.35lb
Caramel 40L	0.22kg	0.5lb

HOPS

Variety	(g/oz)	Add	Attribute
Centennial	35/1.2	Start	Bittering
Centennial	35/1.2	Middle of Boil	Flavor
Centennial	100/3.5	Dry Hop	Aroma

BREWHOUSE ADDITION

None

YEAST

Imperial A62 Bell's House Yeast (recommended)

BIANCA MANGO LASSI GOSE

MODERN GOSE

ABV 6.0% **IBU 15** **OG 1.059**

OMNIPOLLO STOCKHOLM, SWEDEN

Omnipollo are masters of modern fruit-forward beers, and also of beers that evoke memories. Many of their cutting-edge styles result from cofounder Henok Fentie's childhood experiences, or from his trying to re-create them in beer form. Bianca is a traditional Gose expanded to new levels with rock salt, lactose, and mango – as unique as beer, brewing, and art can get.

BASICS

Volume	20L	5gal
Boil Volume	24L	6.3gal
ABV		6.0%
Target FG		1.014
Target OG		1.059
EBC 7 (before fruit addition)		
SRM		3.9
pH		3.5
Attenuation level 73% (before fruit addition)		

METHOD/TIMINGS

MASH TEMP

66°C 151°F 60 mins

FIRST BOIL TIME

15 mins

ACIDIFICATION

pre-acidification to pH ~ 4.5, then ferment with lactobacillus to target pH

SECOND BOIL TIME

60 mins

FERMENTATION

19–21°C 66–70°F

BREWER'S TIP

Add as much mango as possible at the end of fermentation, taking into account that its sugars will become alcohol and the water added will dilute your total ABV. For each 3.5oz (100g) of fresh, raw fruit you're adding about 80% of water and 0.53oz (15g) of sugars.

FOOD PAIRING

A curry of your choice!

INGREDIENTS

Pilsner Malt	3kg	6.6lb
Pale Wheat	3kg	6.6lb

Variety	(g/oz)	Add	Attribute
Magnum	8/0.28	Start	Bittering
Mosaic	35/1.2	End of Boil	Flavor

BREWHOUSE ADDITION

Lactose 480/17 before End of Boil
Sodium Chloride 20/0.71 before End of Boil
Mango (see Brewer's Tip) End of fermentation

Wyeast–London Ale III

STRAIGHT FROM THE TART

SOUR IPA

ABV 6.9% **IBU 35** **OG 1.061**

NEW REALM ATLANTA, GEORGIA

Mitch Steele literally wrote the book on IPA. In 2016, after a decade of being Brewmaster at Stone Brewing, he left and moved to Georgia for a new chapter, cofounding New Realm Brewing Co. Few brewers in the world are better at hop-led pale beers and the twists that can be placed on them, and New Realm's sour series takes IPA to a whole new dimension.

BASICS

Volume	20L	5gal
Boil Volume	27L	7.1gal
ABV		6.9%
Target FG		1.011
Target OG		1.061
EBC		8
SRM		4
pH		3.6
Attenuation level		81.6%

METHOD/TIMINGS

MASH TEMP

67.7°C 153.8°F 60 mins

BOIL TIME

60 mins

FERMENTATION

68°F (20°C) until attenuation, then free rise to 72°F (22°C) for diacetyl rest

FOOD PAIRING

Blackened shrimp and grits, fish and chips

INGREDIENTS

MALT

Pilsner Malt	3.63kg	8lb
Maris Otter	0.70kg	1.5lb
Wheat	0.57kg	1.3lb
Acidulated Malt	0.11kg	0.24lb

HOPS

Variety	(g/oz)	Add	Attribute
No Bittering hop addition.			
Idaho 7	45.4/1.6	Whirlpool	Flavor
Galaxy	45.4/1.6	Whirlpool	Flavor
Ella	45.4/1.6	Whirlpool	Flavor
Idaho 7	47/1.7	Dry Hop	Aroma
Ella	47/1.7	Dry Hop	Aroma
Galaxy	31/1.1	Dry Hop	Aroma

BREWHOUSE ADDITION

None

YEAST

Lallemand Sourvisiae for 67%; Cal Ale yeast for 33% (blend post-fermentation)

BEER PERFECTION BEGINS AT HOME

THE SCIENCE BEHIND THE SUCCESS

Home-brewing may seem like a daunting prospect, at first. When you are knee-deep in packing peanuts from your freshly unwrapped home-brew rig, those chrome pipes and tanks can reflect your worried expression. However, as long as you are armed with a few decent recipes, a drawer full of ingredients, and some weapons-grade sanitizer, you can make a go of it. Start out, learn from your mistakes, and move up that ladder to home-brew greatness.

Also: know this. Even before you begin, there is a buddy in your corner. A way to mitigate those mistakes and spring up that ladder like a Simcoe Superhero. If you can embrace this single overpowering essence, you'll be a far better home-brewer far more quickly, and your bottled or kegged results will bring tears to your eyes rather than your wallet. That force to lean on is this:

Science.

At its most fundamental, brewing is the following of a pathway influenced by science at every turn. The enzymatic reactions inside grain kernels. The action of pH on a solution. Maintaining a healthy and viable colony of microorganisms in a suspension. It's all science. There are many books that focus solely on how you can keep science on your side during your brewdays (see page 217), so let's focus in on a single aspect: off-flavors.

OFF FLAVORS

Those wince-inducing moments when you realize something has gone wrong are to be avoided at all costs. But as you slowly put down the bottle, realize this: something chemical or biological has taken place in your beer, and it can be changed, eliminated, or avoided next time around. Use the science to work in your favor, not against you. Let's take a look at four of the most common off-flavors, and how science can negate their fearsome reputations.

OXIDATION

Oxygen – try living without it. But the same humble gas that fills your lungs every few seconds (don't think about that too much) can be the enemy of beer. Breweries like ours spend untold amounts reducing the amount of oxygen-transfer into beer; purging bottling and canning lines with CO_2 or inert gases to keep it out; measuring levels of dissolved O_2 in developing beer at every stage. We need to know. And if you home-brew, so do you. Because of oxidation.

If you've ever opened a bottle of home-brew and it smells like an old library or dusty bookshop, that's oxidation. Papery aromas, wet cardboard; they all point to the influence of trans-2-nonenal. This unsaturated aldehyde ruins many a batch of beer following oxygen ingress. However, not only is it incredibly hard to brew and also reduce the impact of a gas that is all around us, but yeast needs oxygen to work efficiently.

Yeast cells can respire anaerobically, but oxygen helps them multiply and to produce sterols to bolster their membranes and, in turn, stop them from losing things like acetolactate (see later). Aerating the wort improves attenuation and gets you a better beer. However, too much aeration results in the development of aldehydes that can lead to formation of the dreaded trans-2-nonenal and papery days to come. It's a balancing act, but one that can be resolved thanks to the simplest piece of scientific equipment: a thermometer.

There's a tipping point where that balancing act goes bad – and that's 80°F (27°C). Keep your eyes fixed here and don't aerate your wort if it reaches that mark. Go above that level and you're in danger of creating aldehydes that can't be broken down later. Cool the wort to your fermentation temperature *and then* aerate it for the benefit of your yeasts. They (and a later version of you) will thank you for it.

DIACETYL

Oxidation is also the key cause of another common off-flavor: diacetyl (did we mention that oxygen is the enemy of beer?). If your beer smells like a freshly buttered english muffin topped with popcorn, that's diacetyl. Some diac is OK in certain styles, but not in others, and never in a way that brings to mind a movie theater on a warm evening. However, as its formation involves yeast and oxidation, there are two strikes for science to work on and eliminate it for your follow-up brew.

The key to diacetyl is an amino acid called valine. As yeast cells flurry away in fermentation they need amino acids to grow and reproduce. They can make these themselves or they can go the easy route and absorb them from the liquid they are floating in (which, obviously, is their preferred method). If they make them, then other compounds can result at the same time. As yeast manufactures valine, it also makes acetolactate. And that can be bad news. Some will leak out of the cell and, with oxygen present, is spontaneously converted to the ketone 2,3-butanedione. Diacetyl. Popcorn City.

All is not lost, as brewers know of the stage that brings beer back from the brink: the diacetyl rest. Raising to (if you are fermenting out a lager style) or maintaining (for ales) a fermentation temperature of 64–68°F (18–20°C) for at least 2 days will allow the yeasts to absorb diacetyl back into their tired bodies and convert it to a flavorless compound called acetoin.

But what if the yeasts can't do that? If they are stressed by some other factor, then they may not be able to fully reabsorb the diacetyl. You can buy valine commercially to add into your brew and give the yeast one less thing to create – or you can go a step further and strike back with science.

There now exists an enzyme – acetolactate decarboxylase (ALDC) – that converts acetolactate in the fermenting wort directly to acetoin, bypassing diacetyl completely. A scientific solution in a bottle.

Use if: your typical yeast strain is prone to diacetyl aroma/flavor production.

Alternative: pitch more yeast when packaging and bottle-conditioning your brew. These new yeast cells will work on any diacetyl before you uncap it later.

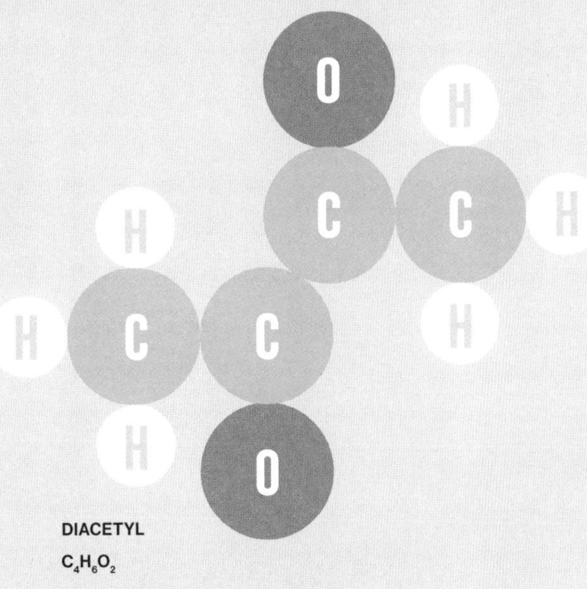

DIACETYL

$C_4H_6O_2$

ACETALDEHYDE

What comes before alcohol? Putting your coat on, usually. But in home-brewing terms (a hobby you may need a coat for, in your garage), the answer is a chemical one. The precursor to alcohol in the pathway that starts with yeast and ends with beer is acetaldehyde. With a characteristic odor of green apples, this compound will eventually be synthesized by yeast into alcohol, and won't be a problem. Usually, that is.

The series of conversions carried out by yeast – and remember, these are single-celled organisms 4/1000ths of a millimeter wide – can sometimes be interrupted. We know they are superheroes, but they are not invulnerable. If yeasts are disturbed, acetaldehyde will remain in the finished beer. If they are super-stressed, they convert the final product (ethanol) *back into* acetaldehyde. And if they die, their cells open and the acetaldehyde leaches out. In short, if you smell your beer and get cider, something has happened to your billions of little helpers.

To ensure that none of this happens and they have a chance to complete their work, you need to maintain a healthy and active colony of yeasts. First of all, you need enough of them to do that work – so pitch a decent amount, particularly if it's a high-ABV beer. Once in, give them time to ferment properly. At least 2–3 weeks at a minimum. Don't take the beer off too early. Keep an eagle eye on the fermentation temperature and don't let it fluctuate.

Yeasts are simple creatures in every sense of the word. They operate in conditions they love and stop when they are faced with even a single hardship. Keep them sweet. Know your culture and respect their rules. Animal husbandry is one of the oldest sciences known to mankind. Although they aren't animals, yeasts are harnessed and fed just like animals during brewing. Look after them, every time you home-brew.

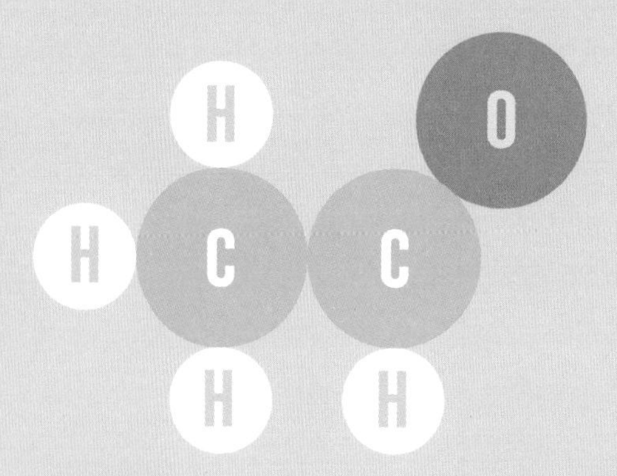

ACETALDEHYDE
CH_3CHO

DIMETHYL SULPHIDE (DMS)

One of the most silvery of linings for every clouded home-brewer is that many of these off-flavors are unstable. When the going gets tough, they don't last long. Many of these unpleasant flavors are fleeting and can be recovered from. Take dimethyl sulphide. DMS gives your home-brew a fairly nasty aroma of canned sweet corn, stewed veggies, or cheap pasta sauce. But it, too, has a weakness.

The key to unlocking that weakness revolves around temperature, and is based on the sole failing of DMS – it is volatile. Your beer can be saved from the rotten flavor of DMS thanks to basic chemistry. If you can remember your old science textbooks, substances that are volatile have higher vapor pressures than others, meaning they are more likely to exist as vapor than liquid at a raised temperature.

As its precursor, S-methylmethionine (SMM), is produced during malt kilning and converted to DMS by the raised temperature of the mash, so too can DMS in turn be converted to another compound by increasing the temperature of the next stage. If you get a vigorous rolling boil going, the volatility of DMS will be its downfall and will vaporize it to dimethyl sulphoxide (DMSO) and into the atmosphere.

There's a caveat to that, though. Be careful if your copper is enclosed as (quite obviously) there's nowhere for the newly vaporized DMS to go, and it condenses on the closed roof and returns to the boiled wort where yeast can convert DMSO back to DMS and your careful planning is all for nought. Leave the lid off (ensuring you have enough head space to avoid a boil-over) and boil for at least 1 hour. Goodbye, dear DMS.

DIMETHYL SULPHIDE (DMS)

$(CH_3)_2S$

HOME-BREW HOW TO...

NEXT STEPS FOR THE ACCOMPLISHED WEEKEND WARRIOR

Ever find yourself idly staring out of the window while you sparge? If you lose focus during even the shorter brewdays, maybe you need a project. Here are five to elevate your home-brew and (maybe) give yourself an even greater grounding into transitioning from your kitchen to a commercial brewery.

HOW TO... BOURBON-BARREL-AGE AT HOME

There are few greater experiences in beer than the first sip of an incredible whiskey-barrel-aged stout. These beers are also one of the true peaks of home-brew lore. And yet, you can create bourbon- or Scotch-whiskey-aged beer at home without a store of dozens of weighty barrels – you just need a pile of oak chips and a willingness to work through the learning curve they present.

Oak chips can be found at any home-brew store, alongside cubes, spirals, or even staves, but chips are the best starting point, and they have massive advantages over acquiring an entire barrel. With chips you always pitch a fresh load, so the full-spirit flavor will carry itself into the beer; barrel flavor will leach out with every new immersion. Chips are always appropriate. Also, they are waaay cheaper.

Adding anything to a beer post-fermentation presents all kinds of pitfalls, but the beauty of oak-chipping a beer is that, if you soak the chips in a jug of whiskey or bourbon for a week beforehand, it both adds flavor *and* sterilizes them. Two for the price of one. The amount of chips to use will vary depending on your rig and the style of beer you are home-barrel-aging, so start slight and work up; around ¾oz (20g) of chips per 5¼ gallons (20 liters) of wort is a fairly safe rule of thumb.

If you pitch them directly into the beer, not only are they hard to remove, but – unless you have really gone to town with soaking them – they will likely float. You want them fully submerged to capture all the flavor, so add them to your beer inside a sterile bag, weighed down with a stainless steel metal spoon, also sterilized beforehand (dunk it in boiling water). Those chips are going nowhere.

Oak chips have a much larger surface area than any other oak-vector, so you get a lot more bang for your buck. This means you can rack the beer off sooner, get it conditioned faster, and drink it more quickly. All of these things are good. You can experiment with different types of oak from different countries, or soaking them in something other than whiskey – why not add a touch of class in the form of a cognac-aged stout? Beats churning out a best bitter every week…

THERE ARE FEW GREATER EXPERIENCES IN BEER THAN THE FIRST SIP OF AN INCREDIBLE WHISKEY-BARREL-AGED STOUT.

BREWDOG

HOW TO...CREATE SOURS AT HOME

Brewers crave cleanliness, purity, and control (or at least, they should). Turning to wild ales and sours is akin to embracing the Dark Side. But, just like the Dark Side, it sure is tempting. The more of these fascinating beers you try, the more it should pique your interest in giving them a try. Is that even possible, though? The answer to this question, when it comes to home-brew, is always yes. And you don't need a doctorate in yeast biology to make them work, either.

The first thing you need to do is think about whether you want to create a beer fermented with brettanomyces or a mixed culture. The first will give you a funky Brett beer, and the second will give you a sour. As you have hopefully read in Chapter 1, Brett beers are not sour. You can re-create the recipe for Funk × Punk on page 182 and that will give you your amazing introduction to the weirdly wonderful world of Brett.

Here, we are going further along the line to a destination beyond the pull of Brett. We are going sour.

These mixed culture beers are fermented out with a combination of lactic acid bacteria – usually (but not exclusively) pediococcus and lactobacillus – and wild yeasts. If you've ever cramped up while working out or reaching for the beer cupboard too quickly, then you'll know about lactic acid. It has a sting to it. Any beer that contains lactic acid bacteria will be sharp, tart, and puckering. A true sour.

So, get yourself some pediococcus and lactobacillus from your usual online supplier of yeast and culture them up. Brew the Funk × Punk recipe again and pitch the lot into your fermentation vessel. Leave it for at least four months – do not move the container during that time and keep sampling to a minimum. You don't want oxygen to damage the carefully evolving flavors. Once you like what you have, transfer it, and there's your first home sour.

YOU DON'T NEED A DOCTORATE IN YEAST BIOLOGY TO MAKE THEM WORK.

You can go a step further and age it on fruit for another three weeks or so (or until the beer has cleared and fruit settled to the base), to give everything an extra dimension. As with all home-brew pro tips, where you take it next is totally up to you. What about a home-soured barrel-aged stout?

HOW TO... BECOME A JACK-OF-ALL-TRADES

We started out brewing beers we wanted to drink, and chances are you did too. Maybe the first few beers were lower-gravity, easy-mash, and simply hopped examples, but once you'd worked out the quirks of your new kit, you likely then moved on to brew whatever beer rocks your socks. Whether it was a series of single-hopped IPAs or a few massively malted imperial stouts, that was when home-brewing became fun.

There's nothing wrong with that, at all. If you really don't care for hefeweizens or whatever, there's no harm in not going near them on your rig. Nothing worse than giving it a shot and ending up with 50 bottles you'd rather not open. Well, unless Christmas is coming up. *"Oh. A dozen bottles of...what did you say it was again?"* But there's a method to the madness of brewing a style you don't like.

First, there's the question of pride. You sat there, thought about it, and came up with that beer. That's going to give it a few more taste points than one you bought from a supermarket. Secondly, tastes change – who's to say you don't actually care for barley wine other than you? Maybe it's been so long that you actually start to dig them again.

A good way to discover and rediscover the joys of the world of beer styles is totry all of them out. The guardians of the beer-style directory, the Beer Judge Certification Program, currently list around 120 different beer styles within their 2015 guidelines (the most recently published). Now, that's a lot of weekends to work through. But there's a way to make it more manageable.

The 120 are broken down into 34 different categories, from "Standard American Beer" that features four sub-sections to everyone's favorite catch-all, the "Speciality Beer". Working your way through the 34, selecting one from each category in turn, is much more doable. This way you can move from American Lager to International Dark Lager, to Czech Pale Lager and Munich Helles.

You could then do Kölsch, Rauchbier, Vienna Lager, Munich Dunkel, Doppelbock, and Weissbier and you've got ten of the most formative global styles in the works in a matter of months. Ahead of you lie the British ales and stouts, Pale Ales and all the fun of the Belgian spectrum. Working through the styles gives you something to aim for, loads of things to enjoy, and, most important of all, it will make you a better brewer as you go.

A GOOD WAY TO DISCOVER AND REDISCOVER THE JOYS OF THE WORLD OF BEER STYLES IS TO TRY ALL OF THEM OUT.

brewdog

HOW TO... SPICE THINGS UP

In the great journey that leads toward home-brewing pro status, many of the wrong turns are caused by extra elements added to your beer. Learning the basics and sticking to recipes that revolve around malt and hops will stand you in good stead, but sooner or later you'll want to spread your wings and try new flavors. Add new things to the pot. And that can lead to a whole world of drainpours.

From the addition of adjuncts like unmalted grains, honeys or fruit to herbs and spices to all manner of other things – anything you include on your brewday will affect the body, pH, and fermentation success as well as increasing the potential for off-flavors, and infection. In short, it can all go wrong in a heartbeat. Even the most fundamental flaw is in play with every batch – the beer simply smells and/or tastes horrible.

This can be chalked up to many things, but primarily the amount and timing of the things you decide to include in the brew. Whether you are adding raspberries, demerara sugar, or nutmeg to your beer (hopefully not all at once), you need to determine how much is enough to add and not overpower everything – or, in regard to the sugar, overprime everything – and when to do the addition.

One way to learn this is to wait right until the end. The best way to practice the effect of including ingredients from your spice cupboard, Asian supermarket, or home-brew ingredients store is to scale everything down. And put the kettle on. Making a "tea" with whatever you are planning to add gives you a micro-scale sample to tinker with and not ruin your big brewday.

Take 1⁄20 of your mash volume of water (so 1¾ pints/1 liter if you have a 5¼-gallon/20-liter batch) and boil it, then stir in a known quantity of your ingredients and leave for 30 minutes. Give it a taste. Can you detect everything in the right proportion? Not too overpowering? If all good, add this flavored "tea" to your priming solution and bottle up. This will give you a known profile to test, and tweak, next time. But this tea-making secret will mean far fewer tweaks in the future.

PUT THE KETTLE ON.

HOW TO...NAIL YOUR DRY-HOP TEMPERATURES

Like pretty much everything else that can be done during a brewday, dry-hopping might seem like a modern trend, but it really isn't. The rise of pithy, resinous West Coast IPAs in the mid to late 1990s pushed the idea of extracting every ounce of bitterness from your hop load, but in truth, brewers have been adding fistfuls of hops into casks before sealing them since the mid-nineteenth century.

Now, of course, our hands merely push the button on our hop cannons, torpedoes, or other ballistic methods of introducing hops to conditioning beer. Dry-hopping is a fact of life, and something we and every other commercial brewery does every day of the week. If you home-brew, chances are you'll dry-hop as well. One of the keys to it, as every home-brewer knows, is how long to leave the hops in there for.

Another key – less discussed but every bit as important – is what the prevailing conditions are for your beer while those hops are doing their thing. The temperature at which you dry-hop is just as formative for your developing beer as the number of days or weeks you leave the hops in contact with the beer for. And, as you can probably appreciate, the warmer you have it, the better.

This is a similar principle to making tea (actual tea, not the "tea" used for flavoring in the previous section). The warmer the water, the more flavor extraction, whether you are brewing tea or beer. The essential oils leaching out of the hops – the entire reason you are dry-hopping – can leach out all the better when it's warmer. So, as you're getting ready to dry-hop, ensure that your ambient temperature is as warm as you can get it and you'll get more bang for your buck.

Ideally, you should aim for the 60–68°F (15–20°C range), so around the ale-yeast fermentation range. Obviously, if you are dry-hopping a lager style, you'll have to raise the ambient to reach this before bringing it down afterward, but the rewards will be worth it (we dry-hop our lagers at around 54°F (12°C), but home-brew-scale can be done warmer than this). Either way, to get the most flavor from your hops, temperature is king.

TO GET THE MOST FLAVOR FROM YOUR HOPS, TEMPERATURE IS KING.

CRAFT BEER FOR THE PEOPLE
Richard Taylor
with James Watt & Martin Dickie

HOME-BREW: FURTHER READING

These days, the internet is the place to go for immediate, ever-changing advice on acing your home-brew adventures. But, as with any other subject, you can't beat a good book. Always there, always able to be held open with a couple of cans, and guaranteed to look good on your shelves, these home-brew bibles will give you bedside reading and instant help when you need it. Embrace the offline life!

Brewing Elements Series (all by Brewers' Publications):

Hops – Stan Hieronymus

Malt – John Mallett

Water – John Palmer and Colin Kaminski

Yeast – Chris White and Jamil Zainasheff

Beer Captured – Tess and Mark Szamatulski (Maltose Pr Llc; 1st edition, 2001)

Brew Like a Monk – Stan Hieronymous (Brewers' Publications, 2005)

Brewing – Ian Hornsey (Royal Society of Chemistry; 2nd edition, 2015)

Brewing Classic Styles – Jamil Zainasheff and John Palmer (Brewers' Publications, 2007)

Designing Great Beers – Ray Daniels (Brewers' Publications, 1996)

Home Brew Beer: Master the Art – Greg Hughes (DK, 2019)

How to Brew – John Palmer (Brewers' Publications; 3rd edition, 2017)

IPA – Mitch Steele (Brewers' Publications, 2012)

Mastering Homebrew – Randy Mosher (Chronicle Books, 2014)

Radical Brewing – Randy Mosher (Brewers' Publications, 2004)

The Complete Joy of Homebrewing – Charlie Papazian (William Morrow; updated edition, 2014)

Wood & Beer; A Brewer's Guide – Dick Cantwell and Peter Brouckaert (Brewers' Publications, 2016)

See also:

How to Build Small Barns and Outbuildings – Monte Burch (Storey Books, 1992)

Home Plumbing Manual (New Ed) – Andy Blackwell (J H Haynes & Co Ltd.; reprint edition, 2014)

How to Clean Just About Anything (Reader's Digest, 2006)

Bookkeeping and Accounting All-in-One For Dummies – Jane E Kelly (John Wiley & Sons, 2015)

Wild Food: A Complete Guide for Foragers – Roger Phillips (Macmillan; Main Market edition, 2014)

GLOSSARY

Acetaldehyde A by-product of fermentation that presents itself as an aroma of green apples.

Alpha acid The source of hop bitterness – a compound found in the resin glands of hop flowers. During the boil it is isomerized (converted by heat) to iso-alpha acids, yielding bitterness in the final beer.

Attenuation The conversion of sugars into alcohol and carbon dioxide by fermentation, reducing the specific gravity of the wort. Beers that are less attenuated will contain more residual sugar and therefore be sweeter than highly attenuated, drier beers.

Black malt Barley roasted at high temperatures in a kiln to give dark coloration and flavors of deep roast and coffee. Often used for stouts or porters.

Brew kettle Also known as the copper. Where wort from the mash is boiled, and hops added.

Chill haze Proteins joining together when a beer is refrigerated, producing particles large enough to cause visible cloudiness when it is poured. Affects the appearance of a beer, not the taste.

Cicerone® Program of learning relating to the serving, culture, and tasting of beer. Participants can take exams for one of four, increasingly complex, levels – Certified Beer Server®, Certified Cicerone®, Advanced Cicerone®, and Master Cicerone®.

Conditioning The maturation of a beer following the fermentation stage. The conditioning phase rounds out the flavors of a beer while preventing the formation of anything unwanted. It also allows the yeast to settle to the bottom of the tank and aids natural carbonation. Can be done at a range of temperatures.

Contract brewing Paying someone else to produce your beer. Often done when a brewery begins life and cannot afford or justify its own kit, or by other breweries that focus on the brand and route to market.

Decoction A technique for mashing grain in which part of the wort is removed, boiled, and returned. This system – developed in continental Europe – raises the temperature of the mash rapidly through a series of these heating steps, resulting in beers with rich, sweet, caramel flavors.

Dextrins Long-chain sugars that yeasts struggle to ferment. Produced by enzymes in barley, they contribute to the gravity, body, and sweetness of the final beer. Lower mashing temperatures produce a higher ratio of dextrin to fermentable sugars.

Diacetyl A powerful compound naturally produced during fermentation that if allowed to remain in the final beer manifests itself as a buttery, butterscotch aroma and flavor. It can be re-absorbed by yeast during the "diacetyl rest" stage.

DMS Dimethyl sulphide: another off-flavor in beer, caused by a low kilning temperature of malt, a short boiling period, or bacterial infection. Results in a characteristic aroma and flavor of cooked sweet corn or tomato soup.

Dry-hopping Addition of dried hops directly to the fermentation vessel to give greater hop aroma and flavor without increasing bitterness. A technique that has become par for the course for craft breweries.

EBC European Brewing Convention: a scale for describing the color of a beer. Similar to the SRM scale.

Enzymes Without enzymes, there would be no beer. They convert starches of malted barley into sugars, which are later used by yeast, and are greatly affected by changes in temperatures and pH.

Esters Compounds created during fermentation that result in fruity, flowery, or spicy aromas and flavors in the final beer. Classically, Belgian beers have high levels of esters.

Fermentation The metabolization of sugars by yeast into ethyl alcohol and carbon dioxide. Or, how beer is born.

Final gravity The specific gravity of a beer when fermentation is complete.

Fining Adding various natural or artificial substances to your conditioning beer to hasten the accumulation and sinking of matter such as yeast cells and produce a clearer beer. Many craft breweries have decided not to use finings, citing differences in flavor or the desire to make vegan-friendly beers (a commonly used fining agent being the fish-derived isinglass).

First runnings The initial wort created during the mash that is high in sugars. Typically added back into the mash (recirculation), historically it was siphoned off and used to make a strong beer, with the rest of the mash being used for a lower-strength version.

Grist Another word for the malt used in brewing, once crushed during the milling process.

Gruit Mixture of herbs used to flavor beer prior to the use of hops.

Heat exchanger Mechanical device used to cool the wort prior to fermentation.

Hop back Vessel used to strain the wort before it is chilled in the heat exchanger. Removes used hops and other debris.

Hot break Boiling of the wort to cause proteins and resins to coalesce quickly.

Hydrometer Instrument used to calculate the specific gravity of a beer. A glass cylinder, the hydrometer bobs up and down in a sample of beer before settling and allowing the gravity to be read from a scale on the side.

Infusion The alternative to decoction, and the most commonly used form of mash. Heated water is added at a specific, single, temperature rather than in a series of increasing steps.

Irish moss Also known as carrageen, a powdered seaweed used to clarify beer by aiding the hot break.

Isinglass Fining agent obtained from the swim bladders of fish and used to clarify a beer during the conditioning stage, accelerating the settling out of yeast to the bottom of the vessel.

Isomerization The change manifested on hop alpha acids during the boiling stage that produces iso-alpha acids, which are soluble and therefore remain in the beer through to the final product.

Kilning The drying of barley to stop the germination phase. Different lengths of kilning time and temperature create malts of different flavors, which yield differing amounts of fermentable sugars.

Krausening Adding a small amount of partly fermented wort from another brew to the conditioning tank. This sparks secondary fermentation and aids the developing carbonation level of the beer.

Lacing The delicate-looking pattern of foam that remains on your finished glass of beer.

Light-struck Exposed to light, UV or fluorescent wavelengths. In beer, these break down the isohumulones in the hops and give a skunk-like, weedy smell.

Liquor In brewing, not the hard stuff; liquor is the name given to water used to mash in.

Maltose Fermentable sugar contained in malt, released in the mash.

Milling Grinding malt to crack the husk and facilitate the release of sugars during the mash. The degree of the crushing will determine how efficient the brew is, and how many grain husks remain to act as a natural filter bed during lautering.

Noble hops Four classic Central European varietals noted for their herbal aromas and flavors. They are Saaz, Tettnang, Spalt, and Hallertau.

Original gravity The specific gravity of a beer before fermentation occurs.

Oxidized Having a stale, cardboard-like aroma and flavor as a result of oxygen acting on the beer.

Phenol Series of aroma and flavor compounds derived from the action of yeast, from the malt or from unwanted bacterial infection. Sometimes they are sought-after by the brewers (such as the clovey aromas of witbiers or the smoky flavor of Rauchbier) and sometimes they are not (the unwanted infections).

Pitching Adding yeast to wort so that fermentation can begin.

Priming Adding sugar to home-brewed bottles to promote secondary fermentation.

Reinheitsgebot German "Purity Law" originating in 1516 and requiring that only malted grains, hops and water be used in the brewing process. Extended to include yeast when it was discovered.

Specific gravity A liquid's density compared to water – essentially, the amount of sugar that is dissolved in a liquid. Pure water has a specific gravity of 1.000.

Top fermentation Caused by yeast cells that remain near the top of the vessel after fermentation. They traditionally work better at warmer temperatures than those that sink to the bottom and produce a beer with a richer, sweeter, fruitier flavor. As such, these yeasts are referred to as "ale yeasts."

Trub Particles left in the bottom of a vessel – the result of allowing a developing beer to rest and have anything in suspension sink to the bottom to be removed. Can be dead yeast, precipitate proteins, hop debris. Nothing tasty.

Volatiles Compounds that evaporate when exposed to air or when in a liquid that is boiled. Some hop oils are notoriously volatile and can be lost during the boil. The volatility of many off-flavor compounds also means that they can be deliberately removed by boiling.

Vorlauf The recirculation of the first runnings of wort – as well as being full of sugars, the initial liquid run-off from the mash will be cloudy and full of particulates, so it is often collected and poured back on top of the grain bed to run through a second time. The process is named after the German word for "temporary."

Whirlpool A vessel that collects the particulate matter following lautering, spinning the wort so that the debris settles in the center and the liquid can be siphoned from the sides.

Wort The proto-beer solution strained from the mash tun. Full of sugars, it is the liquid that yeasts act on to convert it into beer.

Zymurgy The study of yeast and fermentation

COASTLINES
DURRANT DIVED
FRED
2-TALL
WEE IAN
WHERE ARE YOU?!
FRASER
PEACE
Big Suze
Brodie
F.T. CAT
AULD
REYKJAVIK
BREWDOG

INDEX

Specific beers and recipe titles are in *italics, so too are glossary page numbers*

#MashTag 78
21st Amendment Brewery Hell or High Watermelon 91
30 Day IPA 44–5

AB:24: recipe 184
Abbey Ale 58–9
acetaldehyde 36, 204, *218*
acetolactate 202, 203
Aecht Schlenkerla Rauchbier Märzen 137
aeration 202
aging 27, 120, 207
Alaskan Brewing Company 24
Alaskan Brewing Company *Smoked Porter* 95
alcohol and flavor 36
alcohol-free beer 79
aldehydes 202
Allagash Coolship Pêche 32
Allagash Ghoulschip 103
Almanac Loud! 144
alpha acid *218*
American ale yeast 26–7, 29
American Amber 138
American Pale Ale 26
Anchor Mango Wheat 91
Anchor Old Foghorn Barleywine 151
aromas 18. *see also* flavor
attenuation 202, *218*
Augustiner Edelstoff 166
Ayinger Ur-Weisse 152

Baltic Porter 100–1
Bamberg 94
Bamberg Onion, The 130
barley 21, 74–5, 108–9. *see also* malt
barley wine 98–9
Barrel-Aged Imperial Stout 100–1
barrel-aging 27, 120, 207
beer production 23–4, 26–7, 31–2, 202–5
beer styles 82–3, 210
Beersel 55–6
Belgium 110–11
Bell's Brewery 112–13
Bell's Expedition Stout 113
Bell's Porter 112
best before dates 43
Best Bitter (Harvey's Sussex Best) 161
Bianca Mango Lassi Gose: recipe 198
Black Eyed King IMP: recipe 181
black malt *218*
Blind Pig Inaugural Ale 82–3
Bock 106–7
Boon Geuze Mariage Parfait 33
botanicals 102–3
Brasserie De La Senne Zinnebir 164
Brasserie Dupont 110–11
Brett beers 27, 30, 209
Brew by Numbers 07 Witbier Raspberry & Hibiscus 91
brew kettles *218*
BrewDog 12–13, 50, 76, 173
BrewDog Lost Lager 124
BrewDog Vagabond Gluten-free Pale Ale 97
brewing industry 52–3
British food: pairings 139–45, 155–61
Broken Dream: recipe 187
Brooklyn Sorachi Ace 134
Brussels Beer Project Wunder Lager 87
Burning Sky Coolship Release 33
Burnt Mill Steel Cut 97
Burton upon Trent 98
Buxton v Omnipollo Lemon Meringue Ice Cream Pie 93

California 23, 26
Cantillon Gueuze 32
Cantillon Kriek 30
Cascades 61, 62
cheese: food pairings 163–9
chill haze *218*
Chinook 61, 62
Cicerone® *218*
Cigar City Good Gourd 103
Cilurzo, Vinnie 82
Clockwork Tangerine: recipe 177
cold-chaining 40, 44–5
conditioning *218*
contract brewing *218*
Cornell, Martyn 96, 102
Cosmic Crush Tropical: recipe 183
craft beer 7–9, 48–50
Crooked Stave St. Bretta 30
crowdfunding 48–50

dark beer 84–5
Debelder, Armand 54–6
decoction *218*
Deschutes Obsidian Nitro 85
Deschutes the Abyss 101
Devine Rebel (W/Mikkeller) 77
dextrins *218*
diacetyl 36, 203, *218*
dimethyl sulphide (DMS) 205, *218*
dimethyl sulphoxide (DMSO) 205
Disco Soleil: recipe 192
DIY Dog 76, 173
Dogfish Head Punkin Ale 103
DogH: recipe 185
Donzoko Ultrabright 87
Double IPAs 82, 99
Drie Fontenien 55–6
dry-hopping 36, 175, 214, *218*

Einbeck 106–7
Eisbock 106–7
enzymes *218*
esters 27, 36, *218*
European Brewing Convention *218*
Evil Twin Even More Jesus Bourbon Maple Syrup Barrel-Aged 101

Export India Porter: recipe 190

Fallen Brewing Grapevine 29
Far Eastern food: pairings 123–9
Farmageddon Wet Hop IPA 89
Fat Tire 58–9
fermentation 25, 31–2, *218*, *219*
  and flavor 36, 202–5
first runnings *218*
flavor 35–7, 202–5
  pairings 119–20, 131–7
Flensburger Pilsener 133
food. *see* pairing food
Franklin, Sean 60–3
Franklin's Bitter 61
Freigeist Strawberry Eisbock Forever 107
freshness 39–40, 42–3
fruited wheat beer 90–1
Funk×Punk: recipe 182
Fyne Ales Bourbon Barrel-Aged Mills & Hills 101

Gamma Purp 93
Garage Project Hellbender 99
Gas Chromatography–Mass Spectrometry (GC–MS) 37
Germany 104–5, 106–7, 130
gluten-free Pale Ale 96–7
Go-To IPA: recipe 193
Gose 104–5
gravity *218*, *219*
Great Divide Yeti 169
Green Flash Passion Fruit Kicker 91
grist *218*
gruit 102–3, *218*
Gueuze 92–3

Hamilton, Will 108–9
Harviestoun Ola Dubh 12: 101
Hawt DIPA: recipe 196
Hazy Jane 157
Hazy Jane: recipe 176
heat exchanger *218*
herbs 102–3
home brewing 173, 200–5
  beer styles 210
  dry-hopping 214
  recipes 174–99
  sours 209
  and spices 213
  whisky barrel aging 207
hop back *218*
hop-forward pale ales 26, 61, 62, 70
hops 13, 16–18, 61, 62, *219*
  dry-hopping 36, 175, 214, *218*
  and flavor 36–7
  pairings 139–45
  wet hops 89
hot break *218*
hydrometers *219*

Imperial IPAs 82
Imperial Stout 100–1
Imperial Stout Tokyo 97
Inaugural Ale 82–3
India Pale Ale 82–3, 88–9
India Pale Lager 86–7
infusion *219*
intensity 120, 123–9
International Centre for Brewing and Distilling 73–4
Irish moss *219*
isinglass *219*
isomerization *219*

Jackson, Michael 83
Jarl: recipe 186
Juneau 95

Kees American Barley Wine 99
Kernel Brewery 68–71
kettle sour beers 92–3
kilning *219*
krausening *219*
Kuhnhenn Raspberry Eisbock 107

La Trappe Trappist Dubbel 148
lacing *219*
lager 86–7
lambic 54–6, 92
lautering 23–4
Left Hand Milk Stout Nitro 85
Left Handed Giant Life Without Oxygen 99
Lervig Konrad's Stout 129
Lervig Nitro Latte 85
light-struck *219*
Limbach, Lauren 57–9
liquor *219*
Little Earth Project 114–15
Little Earth Project Folly Road 115
Little Earth Project Organic Harvest Saison 114
London 69, 71
Lord Nelson: recipe 188
Lost Lager: recipe 178
The Louisiana Po'Boy 138

Maillard reaction 35
Main Street Fruit Bomb 93
malt 12, 19–21
  chemistry and flavor 35
  pairings 147–53
maltose *219*
Mammoth Fire & Eisbock 107
mashing 23–4, 31–2
Middle Eastern food: pairings 131–7
Mikkeller Peter, Pale and Mary (Gluten-Free) 97
Milk Stout Nitro 85
MilkShake: recipe 189
milling *219*
The Moreton Bay Bug 146

Nanny State 79
New Belgium Brewing Company 58–9
New England IPA (Brewdog Hazy Jane) 157
New Gose - Modern Times Fruitlands 105
nitrogen 85
noble hops *219*

oak chips 207
off flavors 37, 202–5
Omnipollo Bianca Lassi Mango Gose 105
O'Riordain, Evin 68–71
Orval 30, 127
oxidation 202, 203
oxidized 85, *219*

pairing food 119–20
  British 139–45, 155–61
  cheese 163–9
  Far Eastern 123–9
  Middle Eastern 131–7
  sauces 155–61

pairing speciality food
The Bamberg Onion & Smoked Märzen 130
The Louisiana Po'Boy 138
The Moreton Bay Bug 146
Pizza 162
Stornoway Black Pudding 154
Pale Ale 96–7, 146
Palmer, Sir Geoff 72–5
Perrault, Jason 64–7
phenols 36, *219*
pitching *219*
Pizza 162
Pliny the Elder 82–3
porter 84
priming 213, *219*
Pump Action Poet: recipe 180
Pumpkin Ale 102–3
Punk AF 79
Punk IPA 76, 143
recipes 174–5

quality 12–13, 39–40

rauchbier 94–5
recipes
Baked Camembert & Brasserie De La Senne Zinnebir 164
Black Sesame Ice Cream & Lervig Konrad's Stout 129
Carrot Cake & Almanac Loud! 144
Chicken Madras & Brewdog Punk IPA 143
Chinese Stir-Fry with Black Bean Sauce & Scotch Ale (Traquair House Ale) 158
Cobb Salad with Blue Cheese Vinaigrette & La Trappe Trappist Dubbel 148
Five-Alarm Chili & Anchor Old Foghorn Barleywine 151
Flensburger Pilsener & Falafel 133
Grilled Mackerel & Sierra Nevada Pale Ale 140
Harissa Lamb & Aecht Schlenkerla Rauchbier Märzen 137
Key Lime Pie & Ayinger Ur-Weisse 152
Miso Salmon & Orval 127
Smoked Cheese Empanada & Augustiner Edelstoff 166
Spaghetti Bolognese & Best Bitter (Harvey's Sussex Best) 161
Spicy Bean Burger with Barbecue Sauce & New England IPA (Brewdog Hazy Jane) 157
Stilton Rarebit with Walnuts & Great Divide Yeti 169
Sumac Kofte & Brooklyn Sorachi Ace 134
Vegetable Tempura & BrewDog Lost Lager 124
refrigeration 43
Reinheitsgebot *219*
Rhinegeist Wet Hop 89
Rooster's Brewing 62–3
Russian Stout 101

saccharomyces cerevisiae 26–7
Saison 154
Saison Dupont 110
Saison Dupont avec les Bons Voeux 111
sauces: food pairings 155–61
scarification 74–5
Schneider Aventinus Eisbock 107
Scotch Ale (Traquair House Ale) 158
Scotland 108–9, 154
seasonal beers 43
sensory science 59, 119–20
Sierra Nevada Bigfoot 99
Sierra Nevada Brewing Co. 26, 61
Sierra Nevada Northern Hemisphere Harvest Wet Hop IPA 89
Sierra Nevada Otra Vez 105
Sierra Nevada Pale Ale 29, 140
Simcoe 67
Simpsons 20–1
smoked beers 94–5
Smoked Märzen 130
sours 209
Southern Tier Imperial Pumking 103
spices 213
Stone Smoked Porter 95
Stone Tropic of Thunder 87
Stornoway Black Pudding 154
stout 84–5
Straight From The Tart: recipe 199
strong porter 84
Suffolk 114–15
Sumerian Pale Ale 96
Supersonic: recipe 194

Tactical Nuclear Penguin (TNP) 77
Temecula, California 82
Tempest Brewing Loral IPL 87
Thornbridge Tart 93
top fermentation *219*
Top Out Smoked Porter 95
Topcutter IPA: recipe 195
Tröegs Hopback Amber Ale 29
trub *219*
Two Hearted Ale: recipe 197
Twøbays Pale Ale 97

United States 113, 138
food pairings 147–53

valine 203
Vibes Hoppy Pilsner: recipe 191
Victory Harvest Ale 89
Viven Porter 95
volatiles 13, *219*
Vorlauf *219*

water 13, 22–4
Weissbier 90
Westbrook Gose 105
wet-hop IPAs 89
wheat beer 90–1
whirlpool *219*
whisky barrel aging 207
The White Hag the Black Sow Nitro Coffee Milk Stout 85
Witbier 91
wort 31–2, *219*

Yakima Valley 17, 61, 63, 65–7
yeast 13, 25–7
and flavor 36, 202–5
wild yeast 27, 31–3, 209

Zombie Cake: recipe 179
zymurgy *219*

ACKNOWLEDGMENTS

Time to raise a glass to everyone who took the time to help in the production of this book. The beer drinkers of the world thank you!

Leila Alexandre, Darron Anley, Grant Anthony, Carlos de la Barra, Annalena Barrett, Denise Bates, Craig Brown, Megan Brown, Linda Campbell, Stuart Caven, Martyn Cornell, Ben Clark, Armand Debelder, Jamie Delap, Malcolm Downie, Phillip Emerson, Henok Fentie, Amanda Foster, Tom Fozard, Sean Franklin, Paddy Gardiner, Andrew Gibson, Richard Gibson, Allan Grant, Jean-François Gravel, Matthew Grindon, Dave Hall, John Halsall, Will Hamilton, Ben Iddings, Jeppe Jarnit-Bjergsø, Lauren Limbach, Tobias Lund, Rob Mackay, Paul Marshall, Sophie More, Evin O'Riordain, Sir Geoff Palmer, Andy Parker, Jason Perrault, Leah Pilcer, Tim Pritchard, Meghann Quinn, Patrick Robb, Emily Sauter, Simon Shaw, Richard Simpson, Iain Smith, Josh Smith, Kevin Smith, Mitch Steele, Anne & Derek Taylor, Zach Thoren, Lucy Weaver, Lukasz Wiacek, Lizzie Younkin, Dzeti Zait.

PICTURE CREDITS

Photography courtesy BrewDog, except for the following:
3Fonteinen 55. **21st Amendment Brewery** 91. **Alamy Stock Photo** dianajarvisphotography.co.uk 70. **Alaskan Brewing Company** 95. **Bell's Brewery** 112, 113. **Buxton Brewery Co.** 93. **Dogfish Head Craft Brewery** 103. **Elusive Brewing** 188. **Fuerst Wiacek** 196. **Getty Images** David Ryder/Bloomberg 17; Shannon VanRaes/Bloomberg 75. **Left Hand Brewing Company** 85. **New Belgium Brewing** 58. **Omnipollo** Photo by Gustav Karlsson Frost 198. **Paul Winch-Furniss** 125–168. **Richard Clatworthy** 9, 29, 33, 99, 101, 105, 110, 114, 115, 186, 187, 189, 190, 193, 30. **Russian River Brewing** 82. **Saison Dupont** 111. **Schneider Weisse** 107. **Shutterstock** Sergii_Petruk 63. **Sierra Nevada Brewing Co.** 89. **Simpson's Malt** 20. **Stone Brewing - Napa** 23. **Tempest Brewing Co.** 87. **Yakima Chief Ranches** Photo by Digital Vendetta Productions 66.